Numbers

Facts, Figures and Fiction

Richard Phillips

BADSEY
PUBLICATIONS

First edition published in 1994 by the
Press Syndicate of the University of Cambridge.

This second edition published in 2004 by
Badsey Publications, 4 High Street, Badsey,
Evesham WR11 7EW, England.
www.badseypublications.co.uk

Printed by Fuller Davies Ltd, Ipswich, England.

RICHARD PHILLIPS is a freelance educational writer, researcher
and multimedia designer. For 20 years he worked at the Shell
Centre for Mathematical Education at the University of
Nottingham. He has contributed to many educational publications
including *Eureka* (ITMA), *L - A Mathemagical Adventure* (ATM),
Pressure Mat Programs (Panthera), *Coypu* (Shell Centre), *Problem
Pictures* (Badsey Publications) and the award-winning *Population
and Development Database CD-ROM* (Population Concern). He
lives in the village of Badsey in Worcestershire.

ISBN 0-9546562-0-2

Contents

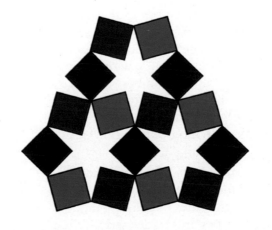

Introduction

First edition

In the opening moments of Peter Greenaway's film *Drowning by Numbers* a girl is seen skipping under a starry sky. She recites as she skips and her huge shadow falls on the farmhouse behind her. Her white dress has a hooped petticoat and is sequined with stars.

'... 98 Rostra, 99 Procyon, 100 Electra.'

She explains she is counting the stars in the sky. When she is asked why she has stopped at 100, she says 'A hundred is enough. Once you have counted a hundred, all the other hundreds are the same.'.

If you really wanted to count the stars you would go on for ever. We all know how to do it, but we also know that the task is impossible. Numbers are at once familiar and ordinary but they are also profound and powerful because they go on for ever and because they make so many things in life possible.

People react to numbers in different ways. For some they are magic or spiritual, but there is also a comic and ridiculous side to them. With counting we have invented something that is bigger than we can comprehend.

This book is about numbers and particularly about a few of the counting numbers. So few, in fact, that its coverage is tiny, or as mathematicians would say, infinitesimal. The numbers from zero to 156 each get an entry, with a briefer mention for all the numbers up to a thousand. But there is almost nothing about the immense column of numbers stretching out beyond this.

In these pages you will find mathematical facts rubbing shoulders with all kinds of other information including the rules of games, bingo calls, film titles and trade marks. Often there is no clear boundary between the mathematics and the other information. Numbers are interesting for many different reasons. Whatever your taste – whether you want to count the stars, count the ways, or just count the spoons – I hope you will find something to appeal to you here.

I would like to thank everyone who has offered ideas and suggestions for this book, particularly John Gillespie, Chris Mills, Judith Mills, Daniel Pead, Thomas Phillips, Ann Shannon and Malcolm Swan. At Cambridge University Press I am especially grateful to Jonathan Barnard, Elizabeth Bowden, Samantha Dumiak, Susan Gardner, Nicholas Judd, Laurice Suess and Rosemary Tennison for their help and support.

Richard Phillips
Shell Centre,
University of Nottingham

Second edition

Ten years on from the first edition, it is fun to be revising the book, writing new articles, producing new illustrations and seeking out photographs. The main entries now span the numbers from zero to 200. It would be tempting fate to claim that I have brought the book up to date. But those who are already familiar with it, should find a little that is new, and hopefully, amusing.

Once again, a big thank you to everyone who has helped. As well as those already mentioned I would particularly like to thank Alan Bell, Rebecca Chaky, Rita Crust, Gerald Jenkins, Elizabeth Noyes and Will Phillips.

Richard Phillips
Badsey, Worcestershire

From 0 to 200

Zero, nought, zilch, nil, nowt...

A *duck* is a score of zero in cricket.

Love is a score of zero in tennis.

Zero is the number of months with 39 days, the number of *g*s in the word *carbuncle* and the number of goal keepers in a game of badminton. It is also the number of cabbages on Mars, the number of hedgehogs who can speak Japanese and the number of fish fingers eaten by William Shakespeare.

Zero is the number of notes in John Cage's piano composition called *4'33"* (4 minutes 33 seconds). The duration of the silence is 273 seconds. This represents the temperature –273° Celsius (called absolute zero). This is the lowest possible temperature, at which molecules have zero heat energy and all molecular motion stops.

Zero is quite a recent invention in the history of mathematics. It had its origins in India in the second century BC. The writings of the Arab mathematician Al-Khwarizmi, who was born around AD680, introduced the Arabic number system and the symbol zero to the West.

Roman numbers (I, II, III, IV, V...) have no symbol for zero and so they are much harder to calculate with than our own. This is because zero allows the use of *place value* in representing numbers. For example, the use of zero allows the symbol *6* to represent six, sixty or six hundred according to its place value.

The rules of arithmetic say you cannot divide any number by zero. Can you see why this is so? Try dividing 6 by 0 on a calculator. What happens? What happens when you divide 6 by numbers which are very close to zero?

(See the end of this book for hints and answers to problems in blue lettering.)

One, unit, unity, single, solo...

An *ace* is number one in playing cards. French playing cards are marked *1* instead of *A*.

A cyclops is a creature with one eye and a dromedary is a camel with only one hump.

There is only one of lots of things. There is only one planet Earth, only one Atlantic Ocean, and there is only one you. All of these are unique.

Words beginning with *uni-* often mean there is *one* of something. For example, unicycles have one wheel and unicorns have one horn. Unisex means the two sexes appearing as one because they are indistinguishable by hair or clothing.

Mono- can also mean that there is *one* of something. A monocle is an eyeglass with only one lens, and a monorail is a railway where the track consists of a single rail. Monochrome means using only one colour, like a black-and-white photograph. Chemical names often include *mono-*; for example carbon monoxide is a poisonous gas whose molecules have only one atom of oxygen.

The letters A, B, C, D, E, M, T, U, V, W and Y all have one line of symmetry.

■ A Möbius strip has one edge and one surface. It is easy to make by taking a long strip of paper, giving it one twist and joining together the ends. Ask one of your friends to colour one side of the strip red and the other side green. This turns out to be impossible because the strip has only one side.

What happens when you split a Möbius strip in two? Carefully cut along the middle of the strip with scissors until you have two separate pieces. What do you notice?

A Möbius strip has only one side.

2 is a prime number and is the only even prime number.

A deuce, a couple, a brace, a duo or a pair...

There are two blades on a pair of scissors and two sides to a piece of paper. People have two hands and so do some clocks. There are two sexes and two sides to an argument. Chess, squash and sumo wrestling are all games played by two competitors. *Two-dimensional* means that something has just length and width, but no depth.

Twain is an Old English word for two. Samuel Langhorne Clemens (1835–1910) worked as a river pilot on the Mississippi. On the boats he often heard a call used for sounding the depth of the water – *mark twain* – which means two fathoms. When he became a writer he adopted this as his pen name. Mark Twain's books include *Tom Sawyer* and *Huckleberry Finn*.

'Two's company, three's a crowd' all depends on who you happen to be with.

'Two heads are better than one' and it may take some brain power to 'put two and two together' but not to know that 'two and two make four'.

If you wait until 'two Fridays come together' you will wait for ever.

There is a saying 'Two attorneys can live in a town, where one cannot'. Lawyers make work for one another.

Two wheels on a bicycle.

Bi- means two. For example, a bicycle has two wheels and a bivalve is an animal with two shells like a mussel.

In binary code numbers are written to the base two. Binary code uses just two digits: 0 and 1. The numbers 1, 2, 3, 4, 5, 6... become 1, 10, 11, 100, 101, 110...

The letters H, I and X all have two lines of symmetry.

Two has a very special property because

$2 + 2 = 4$

and also

$2 \times 2 = 4$.

Try to find a number ? where

$? + ? + ?$

and

$? \times ? \times ?$

both give the same answer. A calculator could help you here.

3 is a prime number.

A triad, triplet, trio, tern or hat-trick...

Tri- means three. So triangles have three sides, tripods have three legs and the dinosaur triceratops had three horns. The French flag is a *tricolore* because it has three colours. Trigonometry is a branch of mathematics based on measuring triangles.

Three-dimensional means that something has length, width and depth.

A tress of hair originally meant a plait or a pigtail with three interwoven strands of hair.

Three-fold symmetry in the tiger flower (Tigridia pavonia).

If the number of petals on a flower is a multiple of three, it is probably from a group of plants called the monocotyledons which includes crocuses, daffodils, tulips, lilies and other plants grown from bulbs.

There are usually three school terms in a year.

Oaths are traditionally repeated three times.

A three-legged race is run by two people each with a leg tied to their partner's.

Billiards is probably the only game played with three balls.

The letters A F H K N Y Z are all made up of three lines.

There are three barleycorns in an inch, three feet in a yard, and three miles in a league. Barleycorns and leagues are some old imperial units of length which are no longer used today.

Once upon a time there were three little pigs ... three billy goats gruff ...
Stories often begin this way and have a similar structure. Number one and number two are always similar so the listener is lulled into believing number three will be the same. But with number three there is a twist in the tale.

In Greek mythology you will find Cerberus, a three-headed dog, and Scylla, a sea monster with six heads. It is curious that mythological heads are inclined to come in multiples of three.

The county of Yorkshire was once divided into a North Riding, a West Riding and an East Riding. *Riding* means a thirding or a third part.

"There are three kinds of lies: lies, damned lies and statistics." declared Benjamin Disraeli (1804 – 1881). As a politician he knew about these things.

With just a ruler and a pair of compasses, it is possible to divide any angle exactly in half. This is called *bisecting* an angle. But is it possible to *trisect* any angle – to divide it in three – using just a ruler and compasses? Hundreds of people have spent hundreds of hours trying to discover a way to do this, without realising that it has been proved to be impossible.

A triangle of triangles in the MOT test sign.

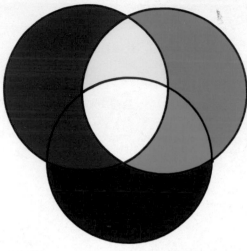

Mixing colours from red, green and blue as on a television.

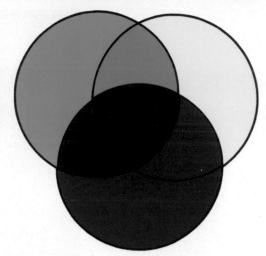

Mixing colours from cyan, magenta and yellow as in the printing of this book.

■ Most colours can be mixed from just three primary colours. But different primary colours are used for different purposes. For example, all the colours you see on a television screen are mixtures of red, green and blue light. With paint you can mix most colours from just red, yellow and blue pigments. The colours in books and magazines are usually printed from three coloured inks: cyan, magenta and yellow, although black ink is used as well.

We use three primary colours because of the way our eyes work. At the back of our eyes are cells called *cones* which are sensitive to coloured light. There are three different types of cone, each sensitive to different wavelengths of light. If our eyes were built differently and we had four types of cone, we would need to use four primary colours in printing, painting and television.

Toblerone chocolate is famous for its triangular packaging. How many triangular chocolate bars would fit into this triangular box? How many would fit into a box one size larger?

4

= 2 x 2

A square number.

A quartet, a foursome...

The word *four* has four letters. In the English language there is no other number whose number of letters is equal to its value.

The number four on a calculator is made up of four light bars.

Many things are arranged in fours. There are four suits in a deck of cards, four points of the compass, and four phases of the moon. There are four wings on a bee and four leaves on a clover, if you are lucky.

The four seasons are spring, summer, autumn and winter. This theme has provided inspiration for many artists, for the composer Vivaldi, and for countless takeaway pizza establishments.

A tetrahedron is a kind of pyramid with four triangular faces. It also has four corners.

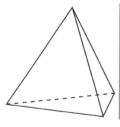 

A tetrahedron and a milk container roughly in the shape of a tetrahedron.

Tetra- means four. A tetradite is someone who attaches mystical properties to the number four. A tetragram is a word with four letters (like *four* itself).

Quad- also means four. A quadruped is a four-footed animal like an aardvark, or almost any animal for that matter.

Four is an unlucky number in China. In the Mandarin language the word for four is also the word for death.

Plus fours are loose baggy trousers which require an extra four inches of cloth in tailoring. This ridiculous male fashion was popular with golfers in the 1920s.

"Four legs good, two legs bad!" chanted the animals in George Orwell's story *Animal Farm*. But one day the ruling pigs had a change of mind and the chant became "Four legs good, two legs *better*!". Orwell was satirising dictators who bend the rules for their own ends.

In a molecule of DNA, just four bases are used to make up the genetic code that determines the distinctive form of every plant and animal. The four bases are called *thymine, adenosine, guanine* and *cytosine,* or just *T, A, G* and *C.*

■ The countries on a map are often shown in different colours. Colours are also used to show states, counties and other regions. To avoid confusion, regions that share a common border are always shown in different colours. What is the smallest number of colours needed to colour the regions on a map? In 1852 Francis Guthrie guessed that the answer is four colours for any map, no matter what shape the regions take.

No one has ever found a map that needs more than four colours. But it has been difficult to find a satisfactory proof that only four colours are needed. In 1976 Wolfgang Haken and Kenneth Appel claimed to have proved the *four-colour conjecture*, but their proof is so complicated, involving hundreds of hours of calculation by a computer, it has been very difficult for other mathematicians to check.

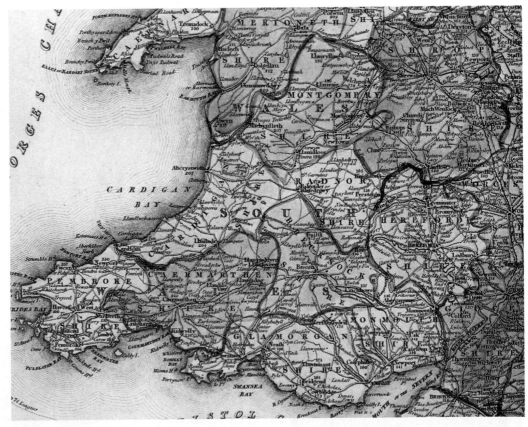

Four colours are all that are needed to colour any map. This map of the counties of England and Wales was published in 1840, just a few years before Francis Guthrie made his famous four colour conjecture.

A quadrilateral is a figure with four sides. But there are several names given to particular types of quadrilateral like squares, rectangles and parallelograms.

Three coins are printed here on the page. Find a similar coin, or any small counter will do.

Place the coin on this page so that the four make the shape of a square. That should be easy.

Now slide your coin around to make a parallelogram. This is not quite so easy.

Now see what other kinds of quadrilateral you can make. Can you make a trapezium? A rectangle? A rhombus? A kite? Any other shapes? Not all of these are possible.

Four-dimensional means that something has an extra dimension as well as length, width and depth. For the scientist, this is usually the dimension of time, where space and time are thought of as part of the same continuum.

However, in mathematics, *four-dimensional* means an imaginary fourth dimension in space. With two dimensions you can draw a square and with three dimensions you can make a cube. With four dimensions it is possible to imagine something called a *hypercube*. Some mathematicians claim to be able to visualise four-dimensional space and can conjure up a clear picture in their heads of a hypercube, which they can rotate or cut in half.

Squares

The first ten square numbers are –
1, 4, 9, 16, 25, 36, 49, 64, 81, 100 ...

Each is the result of multiplying a number by itself –
1 x 1, 2 x 2, 3 x 3, 4 x 4, 5 x 5 ...

– which can also be written –
$1^2, 2^2, 3^2, 4^2, 5^2$ ...

The small '2' means 'squared'.

The square of 7 is 49, and working backwards, we say the square root of 49 is 7.

All square numbers end in either a 0, 1, 4, 5, 6 or a 9. So if someone were to ask you if 225,736,192 was a square number, you could say 'no' right away because it ends in a 2.

If you subtract a square number from the next square number you get the odd numbers 3, 5, 7, 9 ...

1	2	3	4	5	6	7	8	9	10
2	4	6	8	10	12	14	16	18	20
3	6	9	12	15	18	21	24	27	30
4	8	12	16	20	24	28	32	36	40
5	10	15	20	25	30	35	40	45	50
6	12	18	24	30	36	42	48	54	60
7	14	21	28	35	42	49	56	63	70
8	16	24	32	40	48	56	64	72	80
9	18	27	36	45	54	63	72	81	90
10	20	30	40	50	60	70	80	90	100

Square numbers along the diagonal of the multiplication square.

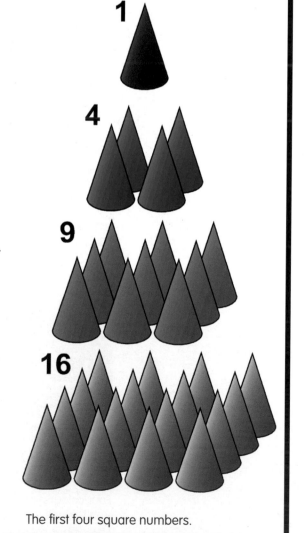

The first four square numbers.

5 is a prime number.

1, 1, 2, 3, 5, 8, 13...
A Fibonacci number.

We have five digits on each hand and foot. V, the Roman symbol for five, may originate from the image of a hand with the fingers spread.

Penta- means five. A pentathlon is an athletics contest with five events and a pentagon is a figure with five sides and five angles. A pentasyllabic word has five syllables, like the word pentasyllabic itself.

Pentagram, *pentangle* and *pentacle* are all names for a five-pointed star. This mystical symbol was supposed to keep away devils and witches. A pentacle headdress folded from fine linen was sometimes worn as a defence against demons.

Many things come in fives: the five senses, the five Chinese elements, and five vowels in the English alphabet.

When things are 'as fine as fivepence' they are pretty good.

A devout follower of Islam worships five times a day facing the holy city of Mecca. The Islamic creed is the 'Five Pillars of the Faith'.

The Five Ks are traditionally worn by The Singhs, who are a brotherhood within the Sikh religion. These are *kes*, long hair; *kangha*, a comb; *kirpan*, a sword; *kachh*, short trousers; and *kara*, a steel bracelet.

Basketball is played with teams of five players, and so is five-a-side football.

The five Olympic rings symbolise the five major continents linked in friendship.

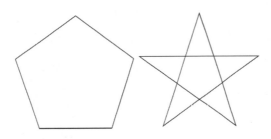

A pentagon (left) and a pentagram (right).

A pentagonal bolt (with five sides) is fitted to many fire hydrants in the USA because it is impossible to undo with a normal spanner – most bolts are hexagonal (with six sides).

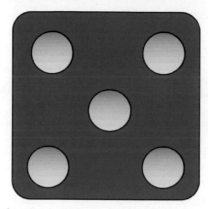

This familiar pattern is called a quincunx.

In Britain, a *fiver* is a five pound note. In the USA, a nickel is a five cent coin.

'Take five!' means take five minute's rest.

The Five Towns, made famous in the stories of Arnold Bennett (1867–1931), are the towns of Stoke, Tunstall, Burslem, Hanley and Longton in the Staffordshire Potteries. Bennett generated some ill feeling in the townspeople of Fenton, who were left out and claimed as much right to inclusion as the other five.

The Five Platonic Solids are the tetrahedron (four sides), the cube (six sides), the octahedron (eight sides), the dodecahedron (12 sides) and the icosahedron (20 sides). These are the only convex regular solids it is possible to construct.

Punch is a drink that traditionally has five ingredients - usually spirits, water, sugar, lemon juice and spice. The name *punch* comes from the Hindi word for five.

> Elizabeth, Elspeth, Betsy and Bess,
> They all went together to seek a bird's nest;
> They found a bird's nest with five eggs in,
> They all took one, and left four in.

This traditional nursery rhyme makes sense when you realise that only one person is involved. Elspeth, Betsy and Bess are variations on the name Elizabeth.

Five was the lucky number for the superstitious French fashion designer Gabrielle 'Coco' Chanel. In 1921 she chose the fifth day of the fifth month to introduce her new brand of perfume which she called *Chanel No. 5.* At that time its scent was unlike any others in a

Five-fold symmetry is found in apples and other plants in the rose family.

Starfish usually have five arms.

market dominated by floral perfumes. It was a huge success and today it is one of the most famous – and most expensive – perfumes available.

Under British law, when you reach the age of five –

* you become 'of compulsory school age',
* you can see a U or PG category film at a cinema,
* you have to pay child's fare on trains,
* you can drink alcohol in private, for example at home.

When you cut through an apple 'the wrong way' you get a five-pointed star. The wild rose has five petals. Apples and roses are part of a large family of

plants called ROSACEAE which includes blackberries, raspberries, strawberries, pears, cherries, plums and peaches, all of which have five-petalled flowers. Although cultivated roses have many more petals, if you look beneath any rose, you will still find five sepals around the base of the flower.

There are only five possible *tetrominoes,* which are shown in the illustration here. A tetromino is a shape made by joining together four squares. *Pentominoes* are shapes made from five squares. How many of these are possible? Only count shapes which cannot be made from each other by flipping them over or rotating them.

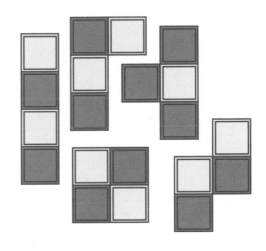

6
= 2 x 3
= 1 + 2 + 3
A triangular number.
= 1 x 2 x 3
Factorial 3 or 3!

The factors of 6 (1, 2 and 3) add up to six. This makes 6 the first perfect number.

1, 2 and 3 make 6 whether you add them together or multiply them.

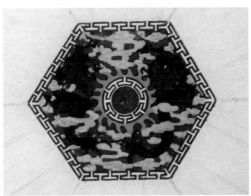

Sex- and *hex-* mean six. So there are six sides on a hexagon and six musicians in a sextet. Sextuplets are six children born together and a hexapod is something with six feet, like an insect.

A cube has six faces and another name for a cube is a *hexahedron*. Six is the highest number on a normal die. An octahedron has six corners or vertices and a tetrahedron has six edges.

Six-legged arthropods include insects like flies, moths, ants, beetles and wasps.

There are six feet in a fathom. A fathom is a unit of length used mainly by sailors. It equals 1.8288 metres.

Volleyball and ice hockey are both played with teams of 6 players. Most morris dances have six dancers.

King Henry VIII had six wives and there are six murder suspects in a game of Cluedo.

'Six feet of earth makes all men equal.' In the past, nearly everyone in England was buried in the ground at a depth of about six feet, no matter what their status.

When you are making a decision it may be 'six of one and half a dozen of the other'.

Hexagonal structures are found in many living things such as the cells of a honeycomb. Carbon, the element that is present in all living matter, has the atomic number six.

In the cult 1967 television series *The Prisoner,* the character played by Patrick McGoohan is called Number Six. He is transported by persons unknown to 'The Village', where everyone is known by just a number.

'Who are you?'
'The new Number Two.'
'Who is Number One?'
'You are Number Six.'
'I am not a number . . . I'm a free man!'
(mocking laughter).

Q – What has six eyes and cannot see?
A – Three blind mice.

'Why, sometimes I've believed as many as six impossible things before breakfast' exclaimed the White Queen in Lewis Carroll's *Alice Through the Looking Glass.*

Above: The hexagonal pattern which forms naturally when metal balls lie on a flat surface.

Below: The hexagonal pattern in a beehive.

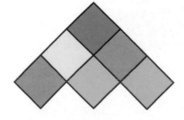

Six as a triangular number.

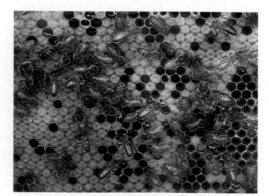

7 is a prime number.

There are seven days in a week.

Hept- or *Sept-* means seven. A heptagon is a figure with seven sides and a heptachord is a seven-stringed musical instrument. A septennium is a period of seven years and September used to be the seventh month in the year, but not any longer.

The Seven Deadly Sins are avarice, envy, gluttony, lust, pride, sloth and wrath (listed in alphabetical order, not order of wickedness).

Among many things that come in sevens are the Seven Wonders of the Ancient World, the Seven Sisters, Shakespeare's Seven Ages of Man, the Seven Levels of Hell, and the Seven Dwarves.

Netball and water polo are both played with teams of seven players.

In Britain the 20p and 50p coins both have seven sides. More precisely, they are Reuleaux figures made from 14 arcs of circles.

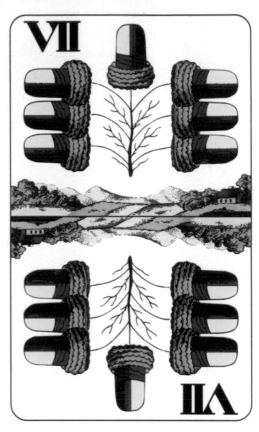

Above: the seven of acorns, a playing card from the Czech Republic.

The Seven Bridges of Königsberg

The city of Königsberg in Germany had seven bridges in the eighteenth century. The people of Königsberg wondered if it was possible to walk through the city crossing each bridge once, and once only. Anyone who tried it found either that they crossed a bridge a second time, or they left a bridge out.

It seemed to be impossible. But they could not be certain it was impossible until the mathematician Leonhard Euler (1707 – 1783) took up the problem. He redrew the map of Königsberg as a *network* (bottom right). The points A, B, C and D represent areas of land and the lines represent the bridges joining them.

Start at one of the points on the network and try to trace a route that goes along each line once, and once only. Can you see why this is impossible?

Euler proved it was impossible by counting the number of lines going to each point. B has five lines, and A, C and D each have three lines. He noted that these were all odd numbers. When he drew a different network with even numbers of lines going to each point, he saw it was possible to trace a route around it. Can you see how this line of thought led him to a proof?

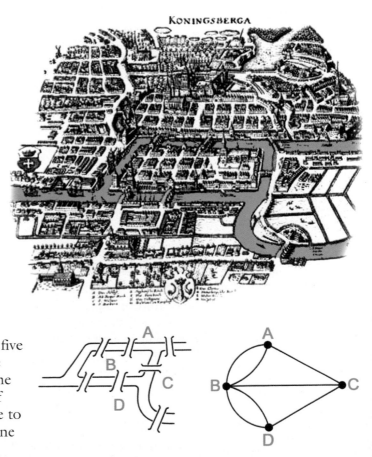

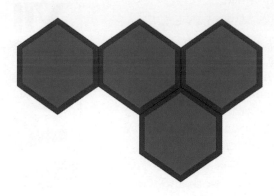

There are seven different ways of linking four hexagons together. Here is one of them. Can you find the other six?

Under British law, when you reach the age of seven you can open and draw money from a National Savings Bank account or a Trustee Savings Bank account.

7-Up is a soft drink. It was invented in America in the 1920s by Mr C L Griggs of Missouri who originally called it *Bib-label Lithiate Lemon-Lime Soda*. With a name like that sales were poor even though the drink tasted good and so Mr Griggs set about changing the name. After six attempts he came up with *7-Up*, or so the story goes. *7-Up* is also the name of a card game.

A *seven shilling piece* was a gold coin in circulation in Britain in the early nineteenth century. Three gold coins were produced: a guinea or 21 shillings, half a guinea or 10 shillings and six old pence, and a third of a guinea or 7 shillings. In modern money seven shillings is 35 pence.

John Sturges's 1960 western *The Magnificent Seven* is about a Mexican village that hires seven gunmen for protection from bandits. The story is based on an earlier Japanese film made in 1954 – Akira Kurosawa's *The Seven Samurai*.

Roy Sullivan, a park ranger from Virginia, USA is the only person to have been struck by lightning seven times. Between 1942 and 1977 he was struck on top of his head (twice), his eyebrows, his shoulder, his chest, his ankle and his big toe. Although he received hospital treatment for his injuries, he was extraordinarily lucky to escape death from so many strikes.

■ Ask a number of different people to give you any number between one and ten, and most will choose seven. Ask people to name their *favourite* number between one and ten, and again most will say seven.

In 1956 George Miller wrote an article *The Magical Number Seven Plus or Minus Two: Some Limits on Our Capacity for Processing Information*. This showed that the amount of information which people can process and remember is often limited to about seven items. One example of this is called the *digit span*.

Ask someone to repeat back to you exactly what you say. Begin with four digits chosen at random e.g. 6 6 2 5. Then give them five digits e.g. 5 8 4 5 0, then six, and so on. Carry on increasing the number of digits until they make a mistake. The longest number of digits they get completely right is called their *digit span* and for most people this is about seven digits.

Above: a wheel with seven spokes.

Below: a Saxon gold brooch found in Worcestershire, England with a pattern repeated seven times.

Suppose someone is shown a pattern of dots for a very short time – just one fifth of a second – and they are asked to count the number of dots they saw. If the number is less than seven they will be right almost every time, but with more than seven, they will make lots of mistakes.

Seven is not really a magic number, but does have an uncanny way of appearing in all sorts of odd situations.

■ A version of this familiar problem appears on the Rhind papyrus written in Egypt about 1650 BC –

As I was going to St Ives,
I met a man with seven wives,
Each wife had seven sacks,
Each sack had seven cats,
Each cat had seven kits:
Kits, cats, sacks and wives,
How many were going to St Ives?

How many were going to St Ives?

8

= 2 x 2 x 2
A cube.

= 2 x 4

1, 1, 2, 3, 5, 8, 13...
A Fibonacci number.

Eight pints make a gallon.

Eight is the third number that stays the same when written upside down.

There are eight legs on a spider, barring accidents. Scorpions also have eight legs.

An eight is a racing boat with eight oars. Its crew is also called *an eight*. There are eight people in a tug-of-war team. A square dance has eight dancers.

Some large car engines have eight cylinders.

According to Indian mythology, the Earth is supported on the backs of eight white elephants.

Before the rise of Christianity, there were eight days in the Greek and Roman weeks.

Pieces of eight go with pirates and parrots. Originally these coins were called *piastre*. They were Spanish silver coins of the seventeenth and eighteenth centuries marked with an '8' because they were worth eight reals. The dollar sign $ is probably derived from the figure '8' as it appeared on 'pieces of eight'.

In the game of pool the *Eight Ball* is a black ball with the number 8. The expression 'behind the eight ball' means to be in a difficult or baffling situation.

A cube has eight corners or vertices.

Many words beginning *oct-* are related to the number eight. An octopus has eight arms and an octet is a group of eight musicians. (An octet of octopuses would therefore have 64 arms between them.)

An octagon is a figure with eight sides and an octahedron has eight faces.

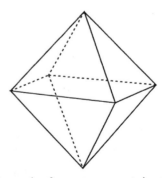

Eight triangular faces on an octahedron.

1 x 8 + 1 = 9
12 x 8 + 2 = 98
123 x 8 + 3 = 987
1,234 x 8 + 4 = 9,876
12,345 x 8 + 5 = 98,765
123,456 x 8 + 6 = 987,654
1,234,567 x 8 + 7 = 9,876,543
12,345,678 x 8 + 8 = 98,765,432
123,456,789 x 8 + 9 = 987,654,321

8-fold symmetry is common in decoration.

← **One Octave** →

An octave is the interval in music between two notes where the higher note has twice the frequency of the lower. On a piano keyboard this corresponds to an interval of eight white notes, for example, the notes C D E F G A B C.

Why is this eight notes rather than seven? We do not say a week has eight days because it starts and ends on a Sunday. Perhaps, to be logical, the interval should be called a *septave* rather than an *octave*? But when you come to play a musical scale it sounds incomplete unless you play all eight notes.

The term *octave* belongs to Western music, but the musical interval occurs in music around the world.

■ The amount of cloud in the sky is measured in oktas on a scale from zero to eight. 8 oktas means the sky is totally clouded and totally clear sky measures 0 oktas. A reading of 6 oktas would mean that three-quarters of the sky is overcast.

Can you place the numbers from 1 to 8 in the cells of this diagram, so that no adjacent numbers are joined by a line?

For example, a line can join the numbers 5 and 8, because they are not adjacent, but the numbers 3 and 4 must not be connected by a line.

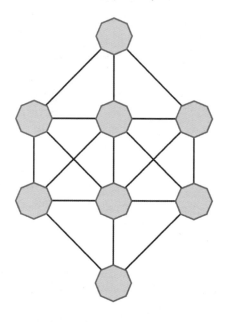

Umbrellas usually have 8 sides, but sometimes have 16.

9
= 3 x 3
A square number.
= 1 + 3 + 5
= 2 + 3 + 4

There are nine major planets in the solar system, Beethoven wrote nine symphonies, and a cat is said to have nine lives.

A polygon with nine angles and nine sides is called a nonagon.

Rounders and baseball are played with teams of nine players.

A game of squash is won by scoring nine points.

Golf courses sometimes have nine holes.

The expression *to the nines* means to the highest degree.

In French the word *neuf* means both *nine* and *new*. In German, the words for nine and new are *neun* and *neu*, and in Spanish, *nueve* and *nuevo*. As you count and reach nine, you know you are about to make a new start.

The game of Shove Ha'penny uses a board divided into nine 'beds' or rows. The winner is the first player to place 3 coins into each bed.

A *nine days' wonder* is something that creates a short-lived sensation.

The 9 of diamonds – the playing card – is sometimes called the *Curse of Scotland*.

Redivider with nine letters is the longest palindromic word in the English language. A palindromic word has the same sequence of letters backwards or forwards.

On cloud nine means happy, euphoric or 'high'. The phrase came into use in the 1950s from a term used by the USA Weather Bureau. For the meteorologists *Cloud Nine* is cumulo-nimbus cloud at a height of 10 km, which is high even by the standard of clouds.

Only about one ninth of the mass of an iceberg is visible above the water. Nearly all its bulk remains hidden beneath the surface.

In the film *2001* the famous black monolith was a cuboid with sides in the ratio 1 : 4 : 9. These are the first three square numbers.

Above and below: Nine-fold patterns on wheels. The bicycle wheel was photographed in China.

■ The game of *skittles* or *ninepin bowling* is many hundreds of years old. The pins are set up in a diamond formation and players throw the ball (or 'cheese') at them. In the nineteenth century some American states passed laws banning the game because bets were often placed on it. But these laws were evaded by the simple ruse of adding a tenth pin. As a result *tenpin bowling* is now the far more popular game.

Over the years, the number of pins used for bowling has varied from as few as three to as many as 17. Can you invent a good way to set up the pins for any number between three and 17?

In this Black Forest game, a spinning top throws coloured balls into nine holes arranged as a nonagon.

The nine cubes on this Swedish stamp make an 'impossible figure'. This arrangement cannot exist in the real world.

■ Here is a game for two players which uses nine coins.

Arrange the coins in a circle. The players take turns and can remove either one coin or two coins which are next to each other. The winner is the player who picks up the last coin.

■ It is easy to work out whether a number is exactly divisible by 9. This is the same as asking whether a number is in the '9 times table'. All you have to do is add up its digits. If the answer is more than one digit long, you add up the digits again, and go on doing this, until you are left with a single digit. If this single digit is 9 then the original number was divisible by 9.

For example, is 781,236 divisible by 9? Adding up its digits –
7 + 8 + 1 + 2 + 3 + 6 = 27
– and adding again –
2 + 7 = 9
Because the answer is 9, the original number must be divisible by 9.

You can also use this method to find out if a number is divisible by 3. If the single digit is a 3, 6 or 9, then the original number was divisible by 3.

This dividing plate is designed to help you cut a round cake or pie into equal pieces. Imagine that you place your cake in the middle of the plate. If you wanted nine pieces you would cut at all the positions marked '9' and also at the position marked '0'. The plate is called a 'plat diviseur' and comes from Brittany in France. The plate lets you divide into 3, 5, 6, 7 or 9 pieces. Why do you think 2, 4 and 8 have been left out?

10
= 2 x 5
= 1 + 2 + 3 + 4
A triangular number.

If a number ends with a zero it is exactly divisible by ten.

We have ten digits on our hands, and ten is the base of our number system: the decimal system. The Roman symbol for ten is X, perhaps representing two crossed hands. A different explanation for its origin is shown below.

The ten pins in a bowling alley are arranged in a triangular pattern.

Virgins, according to the Bible, come in tens: five foolish and five wise.

A dime is a ten cent piece in the USA.

Deca- means ten. So a decade is ten years, a decagon has ten sides and a crab is a decapod because it has ten feet. The Decalogue is a name for the biblical Ten Commandments.

A tithe means a tenth. It was the tax imposed by the church of one tenth of the produce of land and stock.

Men's lacrosse is played with teams of ten players.

Metric measurements use units in multiples of ten. *Deca-* means ten times so ten metres equals one decametre. *Deci-* means one tenth so ten decilitres make a litre.

To ride Bayard of ten toes means the same as *to ride Shanks's pony*. Both expressions mean nothing more than to walk on one's own legs. Bayard was a horse of legendary speed, but Bayard with ten toes means that the only thing that is carrying you is your own pair of feet.

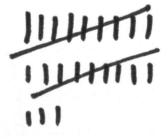

Tally marks are a simple but effective method of counting. Every tenth mark is a diagonal line drawn across the previous nine marks. This convention may be the origin of X, the Roman symbol for ten.

A Czech playing card - the ten of leaves.

■ Marion Walter is a professor of mathematics who has a special interest in the way mathematics is taught. While exploring dynamic geometry software, she discovered a new theorem which has been named *Marion Walter's Theorem*.

Take any triangle and divide each side into three equal parts. Draw lines from these to the opposite corners. A new shape appears in the middle of the triangle – coloured blue in the illustration below. The theorem states that the area of this shape is exactly one tenth of the area of the whole triangle. This works whatever shape of triangle you start with.

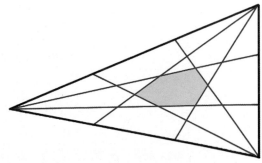

Ten as a triangular number. Can you figure out what the next triangular number is?

What is 10 in Paris and 509 in Rome?

■ Under English law, when you reach the age of ten –

- you can be convicted of a criminal offence if it is proved you knew that what you were doing was wrong (in Scotland this is true from age eight),
- you could be detained 'during Her Majesty's pleasure' for a specific period, including a life sentence, if you are guilty of homicide (murder).

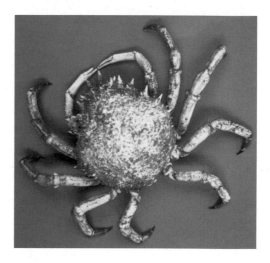

Above: Crabs and other crustaceans like shrimps and lobsters have ten legs.

Below: An unusual drawing instrument that divides a straight line into ten equal lengths.

■ The strength of earthquakes can be measured on the *Richter scale*. The scale measures the amount of energy released by an earthquake. Each point on the Richter scale represents a tenfold increase in energy. So an earthquake with a magnitude of 6 is ten times as strong as an earthquake with a magnitude of 5. The world's strongest earthquakes have magnitudes about 8. Earthquakes measuring between 5 and 6 have been recorded in the UK. The scale was devised by the seismologist Charles Francis Richter (1900 – 1985).

> **This is the size of an A10 sheet of paper.**

■ An A10 sheet of paper is not much bigger than a postage stamp. It is one of the 'A' series of paper sizes. These are cleverly chosen so that the next size can always be made by folding the previous size in half, with the proportions remaining the same.

The largest size is A0 which measures 841 x 1189 mm and has an area of one square metre. It is the size of a large poster. All the other sizes can be easily cut from this without any wastage –
A1 paper measures 595 x 841 mm
A2 paper measures 421 x 595 mm
A3 paper measures 297 x 421 mm
A4 paper measures 210 x 297 mm
  (the size of this book)
A5 paper measures 148 x 210 mm
A6 paper measures 105 x 148 mm
A7 paper measures 74 x 105 mm
A8 paper measures 52 x 74 mm
A9 paper measures 37 x 52 mm
A10 paper measures 26 x 37 mm

Every page has the same width to height ratio which is 1 : 1.4142. The number 1.4142... is called the square root of 2. This means that if you multiply 1.4142... by itself on a calculator you will get an answer very close to 2.

How many A10 sheets can be cut from one A0 sheet of paper?

Packets of printer paper are often marked '80 gsm' This means it weighs 80 grams per square metre. What is the weight of 16 sheets of A4 80 gsm paper?

A passion flower (Passiflora caerulea) has three-, five- and ten-fold symmetry. It has three stigmas, five stamens and ten sepals.

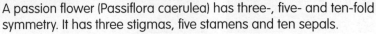

11 is a prime number.

1, 3, 4, 7, 11, 18, 29...
A Lucas number.

The Eleven Plus was an English school selection examination taken by 11 year olds which was abolished in most areas with the introduction of comprehensive schools. It is still fondly remembered by taxi drivers and some education ministers.

Elevenses is a mid-morning refreshment such as tea and biscuits taken around 11am.

Soccer, cricket, American football and field hockey are all played with teams of 11 players on the field.

The sun spot cycle repeats every eleven years. Sun spots are dark areas on the surface of the sun which have strong magnetic fields. When the cycle is at its peak, the number of spots increases causing magnetic storms on Earth which can interfere with radio reception.

The number eleven sometimes appears in the decoration of Easter cakes. The 11 almond paste decorations on this simnel cake represent the 12 apostles, but excluding the treacherous Judas Iscariot.

11 is the fourth number that stays the same when written upside down. The first three are 0, 1 and 8. Can you find the next three? They are all less than 100.

11 marzipan decorations on a simnel cake.

12
= 2 x 2 x 3
= 2 x 6
= 3 x 4
= 2 + 4 + 6
= 3 + 4 + 5

A dozen.

There are 12 months in a year, 12 inches in a foot and 12 old pennies in a shilling.

12 o'clock is midday or midnight. There are 12 signs of the Zodiac, 12 apostles and 12 people in a jury.

12 dancers around a maypole. There are three dancers for each ribbon colour.

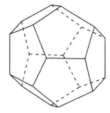

Dodecahedrons have 12 faces.

A dodecahedron is a solid with a regular pentagon on each of its 12 faces. It has 20 vertices (or corners) and 30 edges. If you join together the centres of the faces of a dodecahedron you get an icosahedron.

Both octahedrons and cubes have 12 edges. An icosahedron has 12 vertices or corners.

The wheels of most cars are held on by 16 wheel nuts, with four for each wheel. Two important exceptions to this rule are the Citroën 2CV and the Robin Reliant which each have just 12 wheel nuts. The Citroën 2CV has three wheel nuts on each of its four wheels, while the Robin Reliant has four wheel nuts on each of its three wheels.

There are 12 ways of arranging eight queens on a chessboard so that no queen can capture any other queen (not counting rotations and reflections as being different). Can you find one of these chessboard arrangements?

A cartwheel with 12 spokes. What other symmetry patterns are found on wheels?

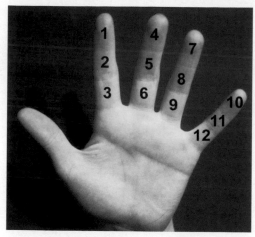

On one hand you can count off the numbers from 1 to 12 by touching the joints of your fingers with your thumb. There are many good arguments for counting in twelves rather than in tens.

Under British law, when you reach the age of 12 –

- you can watch a 12 certificate film,
- you can buy a pet,
- you can be trained to participate in dangerous public entertainments subject to the granting of a local authority licence.

Twelfth Night or the Feast of Epiphany on January 6 marks the end of the Christmas period. It used to be a time of great merrymaking. Shakespeare's play *Twelfth Night* was written to be performed as part of the Twelfth Night celebrations.

In Chinese astrology there is a cycle of 12 lunar years named after animals: the years of the rat, the ox, the tiger, the rabbit, the dragon, the snake, the horse, the goat, the monkey, the rooster, the dog and the pig.

The strength of wind is measured on the *Beaufort scale* which runs from 0 (calm) to 12 (hurricane).

■ Would you buy this 12-inch ruler in a shop?

You may not be impressed by a ruler that has more than half of its marks missing. But it is better than it seems.

You can still use it to measure any whole number of inches from 1 to 12. For example, if you want 2 inches you measure from 10 to 12. For 3 inches measure from 1 to 4, and so on.

It has been proved that you cannot make a 12-inch ruler with fewer marks than this and still measure all the whole numbers of inches.

Design a 6-inch ruler like this which uses as few marks as possible.

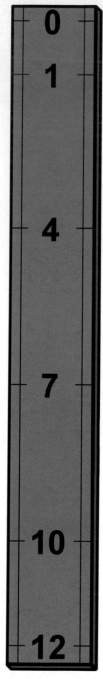

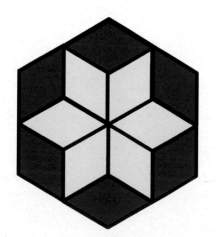

12 parallelograms fit together to make this pattern. What are the internal angles in the parallelograms?

The grating around the base of this tree has 12 sides – it is a dodecagon.

13 is a prime number.

1, 1, 2, 3, 5, 8, 13...
A Fibonacci number.

A baker's, devil's or long dozen.

13 is the 'unlucky number' and Friday the 13th is supposed to be particularly unlucky.

If April 13th is a Friday, how many days will it be to the next Friday the 13th? What is the shortest possible interval between two Friday the 13ths? What is the longest possible interval?

Triskaidekaphobia is the fear of the number 13, for example, of 13 people sitting at a table.

'Eleven plus two' is an anagram of 'twelve plus one'.

Thirteen is the number of hearts in a pack of cards.

Rugby League is played with teams of 13 players.

> When I was but thirteen or so
> I went into a golden land
> Chimborazo, Cotopaxi
> Took me by the hand.
> – W. J. Turner (1889 – 1946)

Under British law, when you reach the age of 13 you can take a part-time job. You can work for up to two hours on a school day or a Sunday. But on Saturdays and in the summer holidays the law allows you to work for up to five hours.

In the sign (above) the sun has 13 rays. The sculpture (below) at Longleat, Wiltshire is a 13 sided figure. It is rare for artists to use 13-fold symmetry, perhaps because their clients may be superstitious.

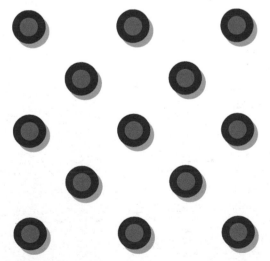

These 13 circles make a centred square. There is a 2 by 2 square enclosed inside a 3 by 3 square. Can you figure out how many circles make the next size of centred square?

I've dealt with numbers all my life, of course, and after a while you begin to feel that each number has a personality of its own. A twelve is very different from a thirteen, for example. Twelve is upright, conscientious, intelligent, whereas thirteen is a loner, a shady character who won't think twice about breaking the law to get what he wants. Eleven is tough, an outdoorsman who likes tramping through woods and scaling mountains; ten is rather simpleminded, a bland figure who always does what he's told; nine is deep and mystical, a Buddha of contemplation...
– Paul Auster, The Music of Chance (1990)

When you buy potatoes in a farm shop, they weigh them with an old pair of scales with weights. Although they only have three weights, they can accurately weigh any whole number of kilograms from 1 kg to 13 kg.

What three weights do they have?

If you look at this picture of a pine cone you can see a pattern of spirals coming out from the centre. They go in two directions. There are 13 spirals going anticlockwise and 8 going clockwise. Both 8 and 13 are Fibonacci numbers. If you count similar spiral patterns in other plants such as sunflowers you can count much larger Fibonacci numbers.

Quite ordinary pine cones show this pattern of 8 and 13 spirals, although it is sometimes easier to see it on the sides of the cone, instead of on the base as in the photograph here.

Fibonacci Numbers

The first ten Fibonacci numbers are –

1, 1, 2, 3, 5, 8, 13, 21, 34, 55...

You calculate Fibonacci numbers by adding together the two previous numbers in the sequence –

$1 + 1 = 2$

$1 + 2 = 3$

$2 + 3 = 5$

$3 + 5 = 8$

$5 + 8 = 13$

– and so on.

Fibonacci numbers are named after Leonardo Fibonacci of Pisa who lived in 13th century Italy. The name is pronounced 'fib - o - na - chee'. His description of the numbers appears in a book called *Liber Abaci*. The numbers have so many interesting mathematical properties that they have become quite famous.

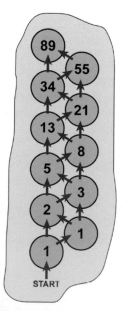

For example, a male bee or *drone* has 1 parent, 2 grandparents, 3 great grandparents, 5 great great grandparents, 8 great great great grandparents, and this continues with the Fibonacci numbers. Although female bees have both a father and a mother, the male drones only have a mother. Try drawing a family tree for a drone and you will see why each generation has a Fibonacci number.

Here is a garden path made of round slabs. As you walk up the path you can step forward on to either of the nearest slabs. The numbers show how many different ways there are to reach each slab. Once again we have Fibonacci numbers.

Squares equal in size to the Fibonacci numbers fit together to make this pattern. As more and more squares are added the shape of the rectangle becomes closer and closer to a number called the Golden Ratio 1.618...

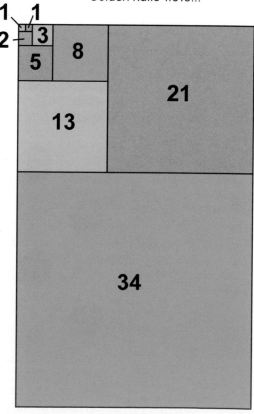

14
= 2 x 7
= 1 + 4 + 9
A pyramidal number – the sum of the first three square numbers.
= 2 + 3 + 4 + 5

There are 14 pounds in a stone and 14 days in a fortnight.

The humble woodlouse has 14 legs arranged as seven pairs.

The flag of Myanmar (formerly Burma) has 14 stars representing its 14 states.

The French word for a fortnight is *quinze jours* or 15 days. If a fortnight begins and ends on a Tuesday, does it contain two Tuesdays or three Tuesdays? And so does it have 14 days or 15 days? Who is being more logical? The French or the English?

Under British law, when you reach the age of 14 –

- you can go into a pub but you cannot buy or drink alcohol there,
- you may be employed on a weekday as a street trader by your parents, subject to local authority byelaws.

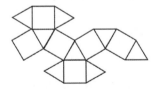

A cuboctahedron is a solid with 14 sides. It has 6 squares and 8 equilateral triangles.

The lower illustration shows one of a pair of gold earrings in the shape of a cuboctahedron. They are about 1500 years old and were found in a tomb in Germany.

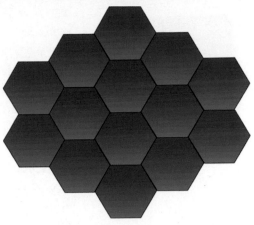

A pattern made with 14 hexagons.

A sonnet is a poem with 14 lines. Shakespeare, Milton, and Wordsworth are among the best known sonnet writers.

■ February 14 is St Valentine's day. The day has its origins in the Roman pagan festival to the god Lupercus as long ago as the the 4th century BC. For the Romans, mid February was a time for courtship. In a lottery, teenage boys drew the names of teenage girls from a box to find their partners for the coming year. In AD 496 the Catholic Church disapproved of this practice and replaced it with their own religious festival celebrating St Valentine. Two hundred years earlier, Bishop Valentine had enraged Emperor Claudius II by marrying couples against the emperor's wishes, had been executed and became a martyr. But the church was only partly successful in removing the romantic associations of St Valentine's day.

The British Museum has a Valentine's Day card sent in 1415 by the Duke of Orleans to his wife while he was imprisoned in the Tower of London. The card shows cupid firing arrows from his bow. The practice of sending cards became more popular, helped by cheaper printing and better postal services. It became possible to exchange cards anonymously and the messages they contained sometimes became more explicit. In the late nineteenth century the Chicago Post Office is said to have rejected 25,000 cards on the grounds they contained messages not fit to be carried through the US mail. In the twentieth century, the Catholic Church downgraded St Valentine and removed him from the calendar of saints' days.

15

= 3 x 5

= 1 + 2 + 3 + 4 + 5

A triangular number.

Rugby Union is played with teams of 15 players.

15 is the constant of a 3 x 3 magic square.

Under British law, when you reach the age of 15 –

• if you are a boy, under certain circumstances you can be sent to prison to await trial,

• you can see a category 15 film,

• you can open a Post Office Girobank account but you must have an adult guarantor,

• you can work up to 8 hours on Saturdays and 35 hours a week in the summer holidays.

The painter and pop artist Andy Warhol (1927 – 1987) predicted 'In the future everyone will be famous for fifteen minutes'. He was jokingly suggesting that everyone should take turns at being famous. What would it mean if we could share out fame? If only one person could be famous at a time, this would allow 96 people to be famous in a 24 hour day and, over a whole year, 35,040 people. But with a world population of about 6 billion, it would take a very long time for everyone to take their turn.

'Fifteen men on a dead man's chest' is one of those lines from sea shanties that don't make much sense. Could it be a commentary on the violent nature of rugby football?

A snooker triangle of 15 balls.

■ '15' on a bottle of sun lotion is its sun protection factor. Sunscreen products have factors like 4, 8, 15 and 20. The higher the number, the more protection from the sun you get.

The number gives you a rough idea of how long you can lie in the sun without burning –

'Safe time' with sunscreen =
   'Safe time' without sunscreen
   x Sun protection factor

For example, if you have fair skin, you may be able to lie without burning for 20 minutes in the hottest midday sun in the UK. But if you wear a sun lotion marked '15' the safe time is roughly 15 times greater –

15 x 20 minutes = 300 minutes

– or five hours. Of course you have to be careful to make sure the sunscreen does not rub off during this time.

The factor number measures how well the sunscreen blocks out the ultraviolet B (UVB) rays of the sun. These rays tan the skin but they also cause burning and are a major cause of skin cancer.

A pattern made from 15 squares.

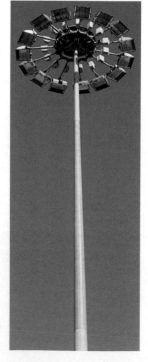

15 floodlights on a pole.

A crystal anniversary celebrates 15 years. The most common cause of celebration is 15 years of marriage, but almost any event can be celebrated in this way, from the reign of a queen, to 15 successful years of trading by a pickle factory. There are named anniversaries for every year from 1 to 15 –

1 year	Cotton
2 years	Paper
3 years	Leather
4 years	Fruit or flowers
5 years	Wood
6 years	Sugar
7 years	Copper or wool
8 years	Bronze or pottery
9 years	Pottery or willow
10 years	Tin
11 years	Steel
12 years	Silk or linen
13 years	Lace
14 years	Ivory
15 years	Crystal

But once we reach 15 years there is a gap and the next recognised anniversary is China at 20 years. It seems sad that once a couple achieves 15 years of marriage, the system assumes that they have nothing really worth celebrating for another five years. With so many wonderful new materials available today to fill the gap, it should not be difficult to put things right. Perhaps the list should continue: 16 PVC, 17 Chipboard, 18 Plutonium...

16
$$= 2 \times 2 \times 2 \times 2$$
A fourth power.
$$= 4 \times 4$$
A square number.
$$= 2 \times 8$$
$$= 1 + 3 + 5 + 7$$

There are 16 ounces in a pound.

Sixteenmo is the size of a book made by folding each sheet of paper into 16 leaves.

In biology, if a cell divides itself in half every 30 minutes, you will have 16 cells in 2 hours.

16 pieces are used by each player in a game of chess.

Caterpillars typically have 16 legs. But when they emerge from their chrysalis as a butterfly or moth, they have only six legs.

16 pebbles feature in Samuel Becket's novel *Molloy* which has one of the longest and most detailed accounts of someone working at a mathematical problem in a work of fiction –

> I took advantage of being at the seaside to lay in a store of sucking-stones. They were pebbles but I call them stones. Yes, on this occasion I laid in a considerable store. I distributed them equally between my four pockets, and sucked them turn and turn about. This raised a problem...

The following six pages of the novel describe different solutions to the problem of ensuring all 16 pebbles are sucked equally often.

Under British law, when you reach the age of 16 –

- you can leave school,
- you can get a full time job, but not in a betting shop or a bar,
- you can have sex,
- you can marry with parental consent (or without it in Scotland),
- if you are convicted of an imprisonable offence, you can be given a community service order,
- if you are a boy, you can join the armed forces with parental consent,
- you can drive an invalid carriage or a moped,
- you can buy cigarettes and tobacco,
- you can have beer, cider or wine with a meal in a restaurant,
- you can apply for a 10-year adult passport.

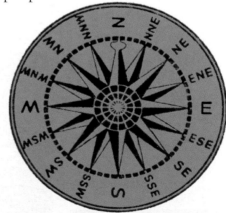

This compass rose is divided into 16 points: north, north-north-east, north-east, east-north-east, east ... and so on. Old maps and charts often included a rose like this to show the compass directions.

17 is a prime number.

In 1963 the Italian film director Federico Fellini released an autobiographical film *Eight and a Half*. A few years later a film called *Seventeen* was released and one film critic, who could not resist making a comparison, headed his newspaper review '*Seventeen* – twice as good as *Eight and a Half*'.

A haiku is a type of poem with just 17 syllables. There is usually a first line of 5 syllables, a second line with 7 syllables, and a final line of 5 syllables. Haikus originated in 16th century Japan but many have been written in English. This example is by the Irish poet Thomas Black –

> hearing the thick hooves
> of a carthorse through the woods
> i forgot my name

East 17 were a successful British boy band in the 1990s. Their name is the postcode for Walthamstow in London.

One of the many achievements of Carl Gauss (1777–1855) was that he discovered a way of accurately drawing a regular heptadecagon – a 17-sided figure – using only a ruler and compasses.

Under British law, when you reach the age of 17 –

* criminal charges against you will be dealt with in the adult courts,
* you can hold a licence to drive most vehicles,
* you can give blood without parental permission.

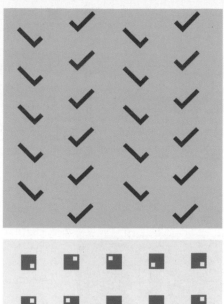

Wallpapers always have a pattern that repeats. But did you know there are 17 different ways in which a pattern can repeat?

■ If you look carefully at wallpaper you will always find that the pattern repeats. You also find repeating patterns on printed fabric. Mathematicians have discovered 17 different types of repeating pattern (called *plane symmetry groups*). Each type is different in the way it repeats – not in the colour or the shape of the pattern. If you had to wallpaper a house for a mathematician, you might use a different one of the 17 repeating patterns on each wall.

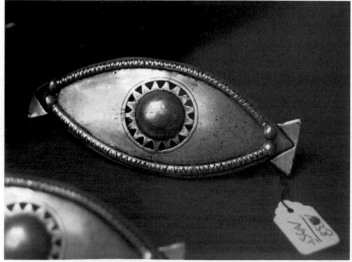

Jewellery with a pattern of 17 holes.

18
= 2 x 3 x 3
= 2 x 9
= 3 x 6
= 3 + 4 + 5 + 6
1, 3, 4, 7, 11, 18, 29...
A Lucas number.

Australian rules football is played with teams of 18 players.

There are 18 holes on many golf courses.

18° Celsius is usually considered a comfortable room temperature.

The two 18 letter words 'conservationalists' and 'conversationalists' are anagrams of each other. They are the longest pair of anagrams in the English language if scientific words are excluded.

The names of racehorses can be no longer than 18 characters including spaces. So if you owned a racehorse you could choose *Chocolate Moonbeam, Gone With The Wind, Pegasus Of The Sky, Personal Guarantee,* or even *Winner of the Race* all of which have exactly 18 characters. But you would not be allowed *Exceeding The Limit* with 19 characters.

Under British law, when you reach the age of 18 –

- you reach the age of majority; you are an adult in the eyes of the law,
- you can vote in elections,
- you can serve on a jury,
- you can buy and drink alcohol in a bar,
- you can open a bank account without a parent's signature,
- you can buy fireworks,
- you can get a tattoo,
- you are entitled to earn the minimum wage for 18-21 year olds.

This Italian poster shows lots of ways of arriving at the answer 18. Can you think of any more?

Eighteenmo is the size of a piece of paper which has been made by cutting a larger sheet of paper into 18 equal pieces. It is also the name for a size of book that has been made in the same way.

A deranged artist always paints her pictures on canvases measuring 6 inches by 3 inches. She believes these are best because their area (6 x 3 =18) is equal to their perimeter (6 + 6 + 3 + 3 = 18). She is thrown into confusion when someone tells her there is another size of canvas where the area equals the perimeter, and where the measurements are in whole numbers of inches. What size is it? Do any other sizes have this property?

6 circles, 12 circles. 18 stars – number patterns on a drain cover.

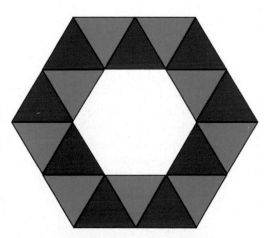

18 triangles make a hexagon ring.

19 is a prime number.

19th Nervous Breakdown was a hit for the Rolling Stones in 1966. It reached No 2 in both the UK and USA charts.

Paul Hardcastle's *19* was a No 1 chart hit in the UK in 1985. Although it was about the American war in Vietnam, the single only reached No 15 in the US charts.

The *Nineteen Propositions* were demands put to Charles I by Parliament in 1642 intended to limit the power of the Crown. The king's rejection of them led to the English Civil War and his execution.

To talk 'nineteen to the dozen' is to talk 'the hind legs off a donkey' or to talk 'a blue streak'.

Someone who is stingy or miserly is said to 'make nineteen bits of a bilberry'. A bilberry is a tiny fruit that would be almost impossible to divide into nineteen parts. There are many similar expressions such as 'He would skin a louse and send the hide to market.'

In the game of cribbage, 19 is an impossible hand.

19 is a *hex number*. These are numbers which can be arranged into a pattern of nested hexagons, as in the illustration. These are the sequence of hex numbers –

7, 19, 37, 61, 91, 127, 169 ...

You will always get a hex number if you take a triangular number, multiply it by six and add one. Hex numbers are also sometimes called *centred hexagonal numbers* but they should never be called just 'hexagonal numbers' as these refer to something else.

This pattern is easy to make with 19 coins. 19 is a hex number.

Countdowns

"Ten, nine, eight, seven, six, five, four, three, two, one, zero. We have ignition!" The launch of a rocket into space begins with a countdown. When the numbers reach zero the rocket leaves the surface of the Earth.

We can count down towards any event in time. In China some traffic lights have a countdown so that everyone knows when the lights are going to change.

Travelling down in a lift or elevator you might see a display showing the floor numbers: 5, 4, 3, 2, 1, 0, -1, -2. The ground floor is shown by '0'. The floors below ground level are '-1' and '-2'. But in some elevators there is no zero and the ground floor is called floor 1. So as you descend the display shows: 6, 5, 4, 3, 2, 1, -1, -2.

There are two types of countdown. One has a zero and the other leaves out the zero. Does it matter which you use?

Suppose you are in a very tall building and you count down ten floors at a time. Using zero the numbers might go: 50, 40, 30, 20, 10, 0, -10, -20 -30. But without zero the numbers would go 50, 40, 30, 20, 10, -1, -11, -21, -31 which is rather messy. Counting down two floors at a time you get 8, 6, 4, 2, 0, -2, -4 if you include zero. But without zero the numbers seem to switch from even to odd: 8, 6, 4, 2, -1, -3, -5 which is much less tidy.

Modern counting systems nearly always include a zero because this gives tidier number patterns which are easier to work with.

Above: after a countdown, liftoff of the Space Shuttle Atlantis in 1994.
Below: a Chinese cyclist watches the traffic lights countdown.

PHOTO: NASA

20
= 2 x 2 x 5
= 2 x 10
= 4 x 5
= 2 + 4 + 6 + 8
= 1 + 3 + 6 + 10

A tetrahedral number – the sum of the first four triangular numbers.

A score is another name for 20. Counting in twenties is common in many languages and cultures. 'Three score years and ten' means 70 years. Marks are sometimes given 'out of 20' and popularity is measured by a 'top 20'.

A china anniversary celebrates 20 years.

In old British coinage 20 shillings made one pound. In old French coinage 20 sous made one franc.

There are 20 hundredweight in 1 ton.

 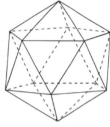

An icosahedron makes a 20 sided dice.

An icosahedron is a solid with an equilateral triangle on each of its 20 faces. It is not possible to make a regular solid with more than 20 faces.

A dodecahedron is a solid with 20 corners (or vertices).

One, two
Buckle my shoe;
Three, four,
Knock at the door;
Five, six,
Pick up sticks;
Seven, eight,
Lay them straight;
Nine, ten,
A big fat hen;
Eleven, twelve,
Dig and delve;
Thirteen, fourteen,
Maids a-courting;
Fifteen, sixteen,
Maids in the kitchen;
Seventeen, eighteen,
Maids in waiting;
Nineteen, twenty,
My plate's empty.

This is a puzzle where you have to arrange the numbers from 1 to 20 in the correct order. The numbers slide around the track and the wheel lets you rotate any four of the numbers.

Would the puzzle be easier or harder if the wheel let you rotate just two numbers? What would happen if you could rotate three numbers?

A darts board is divided into 20 sectors numbered from 1 to 20.

Twenty-twenty vision is perfect eyesight, proved by reading letters on a chart from 20 feet.

Twenty is the name of a village in Lincolnshire near Bourne.

Human infants normally grow 20 teeth, which are replaced by 32 adult teeth as they grow up.

A *hussar* is a soldier on horseback. The word comes from *husz*, the Hungarian for 20. This was the number of households which had to be taxed to provide for one soldier. In the sixteenth century the Hungarian King Mathias Corvinus ruled that each group of 20 households in a town was to provide one soldier on horseback fully equipped.

In Roald Dahl's *Charlie and the Great Glass Elevator* one pill of Wonka-Vite has the effect of making you exactly 20 years younger.

■ *20 Questions* is a game where objects have to be identified by asking up to 20 questions which can only have the answer 'yes' or 'no'.

Have you played the numbers version of *20 Questions*? One person secretly writes down a number and the others ask questions like 'Is it an even number?', 'Is it more than a thousand?', 'Does it have a 6 in it?'.

The game gets harder if you use the rule that each question must be of a different type.

21
= 3 x 7
= 1 + 2 + 3 + 4 + 5 + 6
A triangular number.
1, 1, 2, 3, 5, 8, 13, 21...
A Fibonacci number.

The total number of spots on a normal dice.

Formerly the legal coming-of-age when you received the 'key of the door'.

A game of table tennis is won by the first player reaching 21 points.

Twenty-one is a card game which is also known as *Pontoon* or *Vingt-et-un*. Its object is to have a total score in one's hand nearest to, but not exceeding, 21.

We live in the twenty-first century.

Under British law, when you reach the age of 21 –

* you can become an MP,
* you can hold a licence to drive a large passenger vehicle or heavy goods vehicle,
* you can apply for a licence to sell alcohol.

Onery, twoery,
Ziccery Zan,
Hollow bone, crack a bone,
Ninery, ten.
Spit, spot,
It must be done.
Twiddlum, Twaddlum,
Twenty-one.

In their book *The Lore and Language of School Children*, Iona and Peter Opie describe children's beliefs about the

An unusual grating with 21-fold symmetry.

serial numbers on bus tickets in the 1950s. The most prized tickets had the sum of the digits equal to twenty-one. According to one child, if you found one of these 'you will be lucky all month'. A girl from Ipswich believed that the good luck extended to any multiple of seven: 'If the numbers add up to 7 it means a wish, 14 a kiss, 21 a letter, 28 a parcel.'.

A *twenty-one gun salute* is fired in the UK for royalty and in the USA for the President. It comes from the time when the largest ships of the British navy had 21 guns along one side.

21 is the third star number. Star numbers can be represented by a square with a triangle on each side. Here a 3 x 3 square is surrounded by four triangles.

What is the first star number that is greater than a hundred?

A 'lucky' bus ticket because
3 + 9 + 3 + 6
= 21.

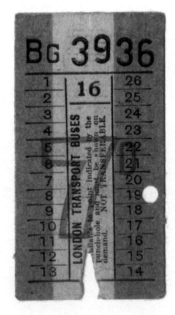

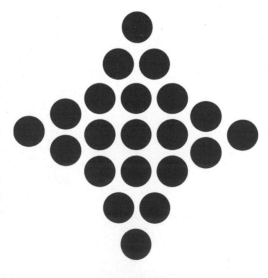

21 as a star number.

22

$= 2 \times 11$

$= 1 + 4 + 7 + 10$

$= 4 + 5 + 6 + 7$

The distance between the wickets on a cricket pitch is 22 yards or one chain. There are many old imperial units of distance like the chain that are only rarely used today.

Snooker is played with 22 balls: a white cue ball, a yellow, a brown, a green, a blue, a pink, a black and fifteen reds.

There are 22 players on the field in a football match.

Bingo calls for 22 include 'Dinkie Doos' and 'Two little ducks'.

■ *Catch-22* (1961) is a book by Joseph Heller which has also been made into a film. It describes the exploits of a group of American airmen based on a Mediterranean island during the Second World War.

The title of the book has become part of the English language. A 'Catch 22' is an impossible situation brought about by two conflicting regulations or conditions. In the book, military regulations say that pilot Orr should be excused from flying further missions because he is insane. But there is a catch –

> Orr was crazy and could be grounded. All he had to do was ask; and as soon as he did, he would no longer be crazy and would have to fly more missions. Orr would be crazy to fly more missions and sane if he didn't, but if he was sane he had to fly them. If he flew them he was crazy and didn't have to; but if he didn't want to he was sane and had to.

Apollo 11 was the first ship to take men to the moon. Its lunar module *Eagle* remained on the surface of the moon for 22 hours while Neil Armstrong and Buzz Aldrin took their historic walk.

Superextraordinarisimo with 22 letters is the longest word in the Spanish language. It means 'extraordinary'.

■ Sometimes it is possible to see a ring around the sun called a *halo*. This is caused by small ice crystals in the sky which bend the sunlight creating the appearance of a ring with a radius of 22°. If you stretch out your arm and look at your closed fist it makes an angle of about 10°. So there are about two fist-widths between the sun and its halo.

The ice crystals in the sky are hexagonal prisms - the same shape as some pencils and ball point pens, but much smaller and made of solid ice. Under certain weather conditions these crystals line up vertically. The halo becomes fainter and instead you see sundogs which are bright patches of light on each side of the sun, also at an angle of about 22°.

There are 22 different ways of linking five hexagons together. Each shape is different and none can be made from another by rotating or reflecting it. The shapes are called pentahexes.

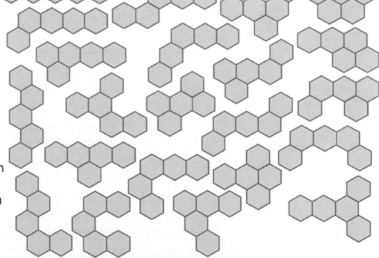

23 is a prime number.

■ When 23 or more people are in the same room there is a better than evens chance that at least two of them share the same birthday. As the average class size in British schools is larger than 23, more than half the classes in Britain have two pupils with the same birthday.

■ 'Twenty-three, skidoo!' is an American catch phrase from the beginning of the century. It may have come from the stage play of Charles Dickens's *A Tale of Two Cities* where an old woman counts the victims being executed on the guillotine. As she speaks 'Twenty-three!' the hero Sydney Carton is beheaded in the last act. Henry Miller's serious dramatisation of the story was parodied by Broadway comedians and became 'Twenty-three, skidoo!'. This is just one of several theories for the origin of the phrase.

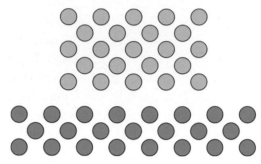

Two different ways of arranging 23 circles.

24
= 2 x 2 x 2 x 3
= 2 x 12
= 3 x 8
= 4 x 6
= 1 x 2 x 3 x 4
Factorial 4 or 4!
= 3 + 5 + 7 + 9

If you had a box of 24 gerbils you could share them out equally among either 1, 2, 3, 4, 6, 8, 12 or 24 people. These numbers are the factors of 24. It is the smallest number to have eight different factors.

There are only 24 hours in a day.

24 furlongs make one league. Furlongs and leagues are old imperial units of distance. A league equals 4.83 km.

In old British money a florin was worth 24 pence. Its modern equivalent is the ten pence piece.

24 is the number of ways in which the letters of the word TEAM may be arranged. Only six arrangements produce words, though.

Advent calendars have doors numbered from 1 to 24. The last door is to be opened on December 24, Christmas Eve.

Twenty-Four Hours from Tulsa was a hit single for Gene Pitney in 1964. It reached No 5 in the UK charts and No 17 in the USA.

Think of any four consecutive numbers (e.g. 392, 393, 394, 395) and multiply them together. The answer will always be divisible by 24.

Above: Nine Men's Morris – the board has 24 holes for pegs.

Below: To win, your ball must fall into one of the 24 cups in this old penny slot machine.

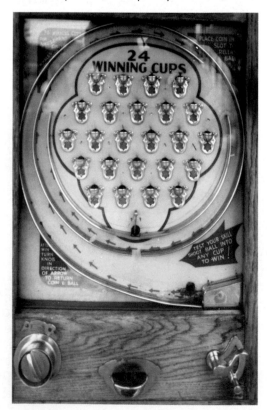

Many traditional songs and rhymes include the phrase 'four and twenty' –

> Four and twenty weavers went out to kill a snail,
> The bravest man among them trod upon his tail;
> The snail turned round with horns like a cow –
> 'God bless us', said the weavers, 'We're dead men now.'

> Sing a song of sixpence,
> A pocket full of rye;
> Four and twenty blackbirds,
> Baked in a pie.

Theories abound about a hidden meaning behind the second rhyme. It has been suggested that the twenty-four blackbirds are the hours of the day, or that the blackbirds are the choirs of monasteries dissolved by King Henry VIII. Another suggests that rhyme is about the printing of the English Bible, the blackbirds being the letters of the alphabet. But the respected authority on nursery rhymes, Iona and Peter Opie, dismiss the need for such explanations. Nursery rhymes get passed on from generation to generation because they are simple and appealing. They are unlikely to have hidden meanings.

24 carat gold means pure gold. 12 carat gold would be 50 per cent pure, 18 carat is 75 per cent pure, and if you are offered 25 carat gold, you are being taken for a ride. The purest form

24 marbles in a 4 x 6 pattern.

of gold used in jewellery is usually 22 carat.

Today if you buy something made of gold it will have a mark showing its purity –
375 means 9 carat gold
583 means 14 carat gold
750 means 18 carat gold
916 means 22 carat gold.

Can you figure out why these 3 digit numbers were chosen?

On this clock how long does the hour hand take to go round? On most clocks it is 12 hours but on this unusual clock it takes 24 hours or a whole day. The clock is at the Greenwich Observatory near London, where it was installed in 1852. It was one of the earliest electrically driven clocks. The hours are shown with Roman numbers. But there is one number the Romans would never have used. What is it?

25
= 5 x 5
A square number.
= 1 + 3 + 5 + 7 + 9
= 3 + 4 + 5 + 6 + 7
= $3^2 + 4^2$
A square number which is the sum of two square numbers.

25 is the sum of the first five odd numbers.

25 per cent means one quarter.

In the USA *a quarter* is a 25 cent piece worth a quarter of a dollar. In the UK, *a pony* is slang for £25.

December 25 is Christmas Day.

A silver anniversary celebrates 25 years.

The M25 is an orbital motorway around London. It is affectionately known as the longest car park in the world.

Pachisi, the Hindu word for 25, is one of the national games of India.

This number square appears on a car park wall in Nimes, France. What is the missing number in the centre shown as N?

Every square number is the sum of two triangular numbers. This pattern shows how 25 can be made by adding 10 and 15. Which two triangular numbers add together to make 100?

This riddle makes sense when you add some punctuation –

Every lady in the land
Has twenty nails upon each hand
Five and twenty on hands and feet
All this is true without deceit

In European countries, television is transmitted at the rate of 25 pictures a second or 90,000 pictures in a one hour programme. In the cinema, films are usually projected at a slower speed of 24 pictures a second.

The film *Jurassic Park* consists of about 182,880 separate pictures. How long does the film last in the cinema at 24 pictures a second? How much shorter would it be if it was shown on television at 25 pictures a second?

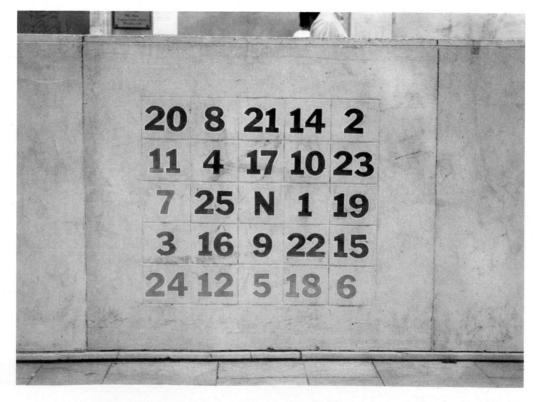

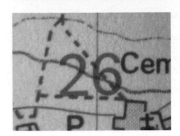

26

= 2 x 13

= 5 + 6 + 7 + 8

There are 26 letters in the English alphabet.

Precipitevolissimevolmente with 26 letters is the longest word in the Italian language. It means 'as fast as possible'.

December 26 is Boxing Day (except when December 26 is a Sunday).

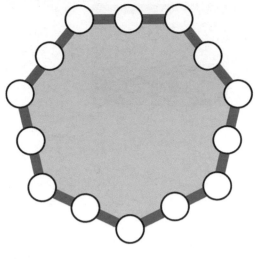

This net makes an unusual solid with 26 faces and 24 corners. Can you imagine what it looks like?

Henry Dudeney (1847 – 1930) was an ingenious inventor of mathematical problems. This is his Heptagon Puzzle which is quite difficult to solve.

Using the numbers from 1 to 14 place a different number in every circle so that the three numbers along every side add up to 26.

27

= 3 x 3 x 3

A cube.

If you add up all the numbers between 2 and 7 the total is 27.

The coloured balls in snooker have a total value of 27.

27 is the international telephone dialling code for South Africa.

There are 27 books in the New Testament of the Bible.

Three-dimensional noughts and crosses is played on a 3 x 3 x 3 grid giving 27 positions to place your nought or cross.

This wooden cube has been painted red and cut into 27 smaller cubes.

If you take it apart, how many of the smaller cubes will have no red paint on them? How many will have one face painted red? How many will have two faces painted red? How many will have three faces painted red? How many will have four faces painted red?

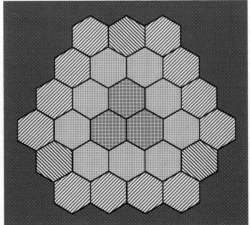

A patchwork design using 27 hexagons.

28
= 2 x 2 x 7
= 2 x 14
= 4 x 7
= 1 + 2 + 3 + 4 + 5 + 6 + 7
A triangular number.
= 1 + 2 + 4 + 7 + 14

28 is a perfect number because it equals the sum of all its factors.

There are 28 days in February, except in leap years.

Sets of dominoes usually have 28 tiles.

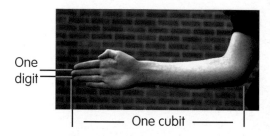

One digit

One cubit

There are 28 digits in one cubit. The ancient Egyptians had their own system for measuring length based on body lengths. A digit was based on the width of one finger (about 19 mm). A cubit was the distance measured from the elbow to the tips of the fingers, and this equalled 28 digits.

Use a ruler to measure a digit and a cubit on yourself. How many of your digits make one cubit? How close is it to 28? Take measurements from a number of different people. How close are they to 28?

Perfect numbers

Perfect numbers are quite rare. The first five are –
6, 28, 496, 8128, 33550336 ...
The tenth perfect number is more than 50 digits long.

A perfect number is defined as a number that equals the sum of its factors, apart from itself. The factors of 28 are 1, 2, 4, 7 and 14, adding these up makes 28, and so 28 is a perfect number.

Perfect numbers have been known for many thousands of years and have often been treated with reverence. Jewish and Christian writers stressed that the world was created in six days and that there are 28 days in a lunar month.

All known perfect numbers are even numbers and are triangular numbers. They all end with either a '6' or an '8'.

29 is a prime number.

1, 3, 4, 7, 11, 18, 29...
A Lucas number.

There are 29 days in February in a leap year.

All the months in the Jewish calendar and the Muslim calendar have either 29 or 30 days.

29 is the highest possible hand in the game of cribbage.

29% of the surface area of the earth is land.

Seven straight cuts through a pizza can divide it into as many as 29 pieces.

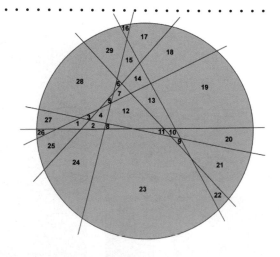

One way to divide a pizza into 29 pieces with 7 straight cuts.

In the stereotypes loved by comedians, 'Lot 29' was to auctions what 'PC 49' was to policemen. 'Lot 29' was likely to be a hideously ugly vase, a chair on the point of falling to pieces, a radio that only crackled, or some other object that no one in their right mind would want to buy.

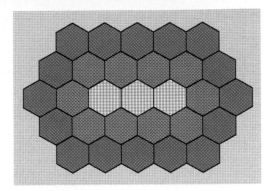

29 hexagons in this patchwork pattern.

30
= 2 x 3 x 5
= 2 x 15
= 3 x 10
= 5 x 6
= 2 + 4 + 6 + 8 + 10
= 4 + 5 + 6 + 7 + 8
= 1 + 4 + 9 + 16

A pyramidal number – the sum of the first four square numbers.

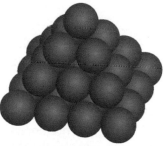

30 is a pyramidal number. The base of this square is a 4 x 4 square with smaller squares resting on it.

There are 30 days in April, June, September and November.

Dodecahedrons and icosahedrons both have 30 edges.

30 mph is the UK speed limit for vehicles driving in a built up area.

Thirty something was a television drama series about American yuppies in their thirties.

The Thirty Years' War was a European war from 1618 to 1648.

In old British money half-a-crown (or half-a-dollar) was a coin worth 30 old pence.

A pearl anniversary celebrates 30 years.

The best known feature of Stonehenge is the Sarsen Circle which was built as 30 upright stones with 30 lintel stones on top of them. Only 16 remain standing today.

30 is the area of a right-angled triangle with sides of 5, 12 and 13. It also has a perimeter of 30. Can you find a right-angled triangle which has an area and a perimeter both equal to 24?

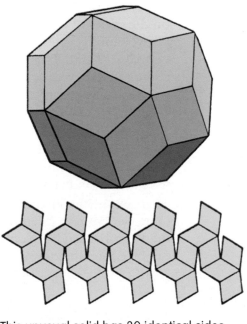

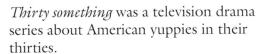

Eggs are usually sold in multiples of 6. Here there are 30.

This unusual solid has 30 identical sides which are rhombuses. It is called a rhombic triacontahedron.

31 is a prime number.

$= 2^0 + 2^1 + 2^2 + 2^3 + 2^4$
The sum of five powers of 2.

$= 5^0 + 5^1 + 5^2$
The sum of three powers of 5.

$= 2^5 - 1$
A Mersenne prime number.

There are 31 letters in the Cyrillic (Russian) alphabet.

There are 31 days in most months.

31 is the international telephone dialling code for the Netherlands.

Thirty-one is a betting game played with cards.

In French the expression *trente et un* means someone who is well dressed.

Days in a Month

The month is a convenient way to measure time based on the interval from one full moon to the next. When measured accurately this is 29.53 days and because this is not an exact number of days, people have devised calendar months which vary in length in order to fit neatly into the calendar year.

The Muslim calendar is closely based on the cycles of the moon. It has twelve months which alternate between 29 and 30 days in length, adding up to a year of 354 days. This results in a calendar which rotates around the seasons.

The Jewish calendar also has months of 29 and 30 days, but by allowing either 12 or 13 months in the year, the seasons remain fixed.

There are many other calendar systems. Our own Western calendar, based on the Roman calendar, has months which can be either 28, 29, 30 or 31 days.

In 1577 William Harrison gave this familiar way to remember the length of each month –

> Thirty dayes hath November,
> Aprill, Iune and September;
> Twentie and eyght hath February alone,
> And all the rest thirty and one,
> But in the leape you must adde one.

There are many variations on this mnemonic rhyme; some are more helpful than others –

> Dirty days hath September,
> April, June and November.
> From January up to May
> The rain it raineth every day.
> February hath twenty eight alone,
> And all the rest have thirty-one.
> If any of them had two and thirty
> They'd be just as wet and dirty.
>   – Tom Hood

The interval of 29.53 days between full moons is called the *synodic* month. A *sidereal* month is the time for the moon to complete one orbit around the earth and equals 27.32 days.

32
= 2 x 2 x 2 x 2 x 2
The fifth power of 2.

= 2 x 16
= 4 x 8

Take a sheet of paper and fold it in half. Fold it in half again and go on folding until you have made five folds altogether. Open it up and you find 32 rectangles on the sheet of paper. One extra fold would make 64 rectangles. It has been claimed that it is impossible to fold any sheet of paper more than seven times. This was tested on a Canadian TV programme. A piece of paper measuring 100 yards square was laid out on a football pitch. They made eight folds without too much difficulty, but could only maintain a ninth fold by sitting on the paper.

32° Fahrenheit is the melting point of ice.

Traditionally there are 32 points of the compass.

Adults normally have 32 teeth: two incisors, one canine, two premolars and three molars on each side of each jaw.

■ A football is usually made by sewing together 32 panels – 12 regular pentagons and 20 regular hexagons. The shape, which makes a good approximation to a sphere, is called a truncated icosahedron. It has 32 sides and 60 corners. It is the shape you get by taking an icosahedron and cutting off each of its corners.

The same shape occurs in the molecule of buckminsterfullerene (C_{60}) which is a new form of carbon discovered in the 1980s. Each molecule has 60 carbon atoms arranged at the corners of a truncated icosahedron; it is the most spherical molecule known to science and a purer form of carbon than either graphite or diamond. This orange-coloured material has many remarkable properties. Other atoms and molecules can be trapped inside its cage-like molecular structure.

The football shape – a truncated icosahedron.

Jeu de 32 cartes. This French pack of playing cards contains only 32 cards. Which cards have been left out?

33
= 3 x 11
= 3 + 4 + 5 + 6 + 7 + 8
= 9 + 11 + 13
= 1! + 2! + 3! + 4!
The sum of the first four factorials – and in the ASCII code used by computers, 33 represents an exclamation mark.

English solitaire boards are circular with 33 hollows to take marbles.

Bingo calls for 33 include 'All the threes' and 'Fevvers!' which comes from the Cockney tongue twister 'firty-free-fahsand fevvers on a frush's froat' (33,000 feathers on a thrush's throat).

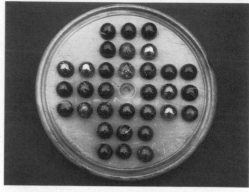

An English solitaire board with 33 hollows.

33 is the international telephone dialling code for France.

The writer Oscar Wilde published under the pseudonym C33 because he occupied cell 33 at Reading Gaol.

34
= 2 x 17
= 7 + 8 + 9 + 10
1, 1, 2, 3, 5, 8, 13, 21, 34 ...
A Fibonacci number.

34 is the international telephone dialling code for Spain.

■ All 4 x 4 magic squares have the constant 34.

Although there are many different 4 x 4 magic squares, the example shown here is particularly special as it is a *diabolic magic square*.

Add up any row or column and the total is 34. Total any of the diagonals, including broken diagonals like 4 + 10 + 13 + 7 and the answer is still

A diabolic magic square.

always 34. Take any group of four cells and these add up to 34, including the four corner cells.

35
= 5 x 7
= 2 + 3 + 4 + 5 + 6 + 7 + 8
=1 + 3 + 6 + 10 + 15
A tetrahedral number – the sum of the first five triangular numbers.

35 mm film is a popular format for modern cameras.

Thirty-five or *Trentacinque* is an Italian card game where bets are placed in a central pool and won by the player who holds cards in one suit to the value of 35 or more.

A coral anniversary celebrates 35 years.

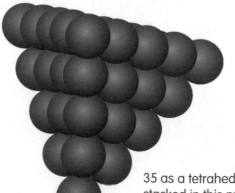

35 as a tetrahedral number. Piles of cannonballs were sometimes stacked in this pyramid shape built up in triangular layers. The sequence of tetrahedral numbers is 1, 4, 10, 20, 35, 56, 84, 120...

A tandem bicycle usually takes only two people. In 1979 *Pedaalstompers Westmalle* of Belgium built a tandem cycle for 35 riders. It measured over 20 m long.

'Thirty five years' was a catchphrase popularised by the comedian Kenneth Williams. Each episode of the 1960s BBC radio programme *Beyond Our Ken* included a quavery old man played by Williams. When asked how long he had been doing anything, he always answered 'Thirty five years, I been in this job *thirty five years*.'

The sign means that there is a steep climb ahead. What does the 35% mean? Would a 40% hill be more steep or less steep?

36
= 2 x 2 x 3 x 3
= 2 x 18
= 3 x 12
= 4 x 9
= 6 x 6
A square.

= 1 + 2 + 3 + 4 + 5 + 6 + 7 + 8
A triangular number.

There are not many numbers which are both square and triangular. The first four are –

1, 36, 1225, 41616 ...

$= 1^3 + 2^3 + 3^3$
The sum of the first three cubes.

36 is the smallest number to have nine different factors. These are 1, 2, 3, 4, 6, 9, 12, 18 and 36.

36 is the international telephone dialling code for Hungary.

There are 36 inches in one yard.

Two dice can fall in 36 different ways.

William Shakespeare wrote 36 plays.

Thirty-six is a game played with one die. Players take turns to throw the die or pass. The winner is the person with the score closest to 36, but anyone with a score of more than 36 is out.

A popular size of photographic film has 36 exposures.

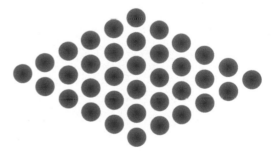

A pattern of 36 – any square number can be arranged like this.

A market researcher knocks on a door which is opened by a woman. 'How many children do you have?' he demands. 'Three' she answers. 'How old are they?' The woman considers this a bit nosey and so she says 'If you multiply their ages together the answer is 36. Look at the number on the house next door. That is the sum of their ages.' The market researcher is not put off by this and says 'I still need one more clue' so the woman says 'My oldest plays the piano. Now go away.' The market researcher now knows the ages of the three children. How old are they?

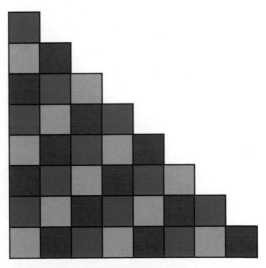

36 as a triangular number.

These 36 stamps are arranged as a 4 by 9 block. By tearing them into two pieces, it is easy to rearrange them as a 2 by 18 block. But can you see a way off tearing the same 4 by 9 block into two pieces and rearranging them as a 6 by 6 block?

37 is a prime number.

37° Celsius is the normal body temperature.

The French version of solitaire uses an octagonal board with holes for 37 pegs. The English solitaire board has only 33 holes.

37 x 3 = 111
37 x 6 = 222
37 x 9 = 333
37 x 12 = 444

What should you multiply 37 by to make 999?

Other multiples of 37 have curious properties. For example 148, 481 and 814 are all multiples of 37. All three numbers use the same digits. The same is also true of –

185, 518 and 851
259, 592 and 925
296, 629 and 962

– all of which are multiples of 37.

The number has yet more odd properties –

$37 \times (3 + 7) = 3^3 + 7^3$
$3^2 + 7^2 - (3 \times 7) = 37$

As a decimal 1/37 has a simple pattern of three repeating digits –
0. 027 027 027 ...

37 as a hex number.

Here the hexagons are printed in two colours. Try to find another way of colouring them so that every hexagon has a different colour from those next to it. What is the least number of colours needed to do this?

38

= 2 x 19

= 8 + 9 + 10 + 11

In Hong Kong 38 is a lucky number and car registrations like 3838 are much in demand from business people. In Cantonese the word for three is pronounced 'sam' which is also the Cantonese word for life. Similarly, eight is pronounced 'fa' which is also the word for wealth. So lucky 38 should bring you a long and wealthy life.

38°C is the highest air temperature ever recorded in the UK. Brogdale in Kent recorded 38.5°C (101.3°F) on Sunday 10 August 2003.

The thirty-eighth parallel is the border which divides North and South Korea. In 1945 at the end of the World War II a line was drawn on a map along the line of latitude to divide the countries.

■ Every row of numbers in this magic hexagon adds up to the same total – 38. There are 15 different ways you can make 38.

This pattern was discovered by an American railway clerk named Clifford Adams. Without knowing whether it was possible to make any magic hexagon, he began his search in 1910. He used a set of ceramic hexagon tiles marked with the numbers from 1 to 19. Again and again he tried arranging these in different ways so that all the

The only possible magic hexagon.

rows added up to the same number. He worked at the problem on and off for 47 years before discovering this arrangement while recovering from an operation.

Unfortunately he then lost the piece of paper on which he had written the solution. He was unable either to remember it or reconstruct it. But five years later in December 1962 he found the missing piece of paper and sent his magic hexagon to the mathematics writer Martin Gardner. Gardner passed it on to Charles Trigg who revealed that no one had ever discovered a magic hexagon before. Furthermore Trigg was able to prove that no other magic hexagon of any size was possible.

• •

39

= 3 x 13

= 4 + 5 + 6 + 7 + 8 + 9

= 11 + 13 + 15

39 is the international telephone dialling code for Italy.

The Thirty-Nine Steps is a spy story by John Buchan which has been made into several films, of which the best is thought to be Alfred Hitchcock's 1935 version which stars Robert Donat and Madeleine Carroll.

The 39 Articles are the constitution of the Church of England. There are 39 books in the Old Testament of the Bible.

There is a children's riddle that goes 'I have four tea cups and I break one. How many are left?' 'Three' the victim answers 'No, there are 39 left. I said I

had forty cups and broke one.'

For the mathematician, 39 does not have much to offer in interesting properties. In the first edition of his *Dictionary of Curious and Interesting Numbers* David Wells claimed that 39 '...appears to be the first uninteresting number, which of course makes it an especially interesting number ... It is therefore the first number to be simultaneously interesting and uninteresting.' Inevitably this statement produced a series of claims for mathematically interesting properties for 39 and David Wells was forced to change the entry in the second edition of his book. This lists 51 as the first uninteresting number. No doubt, there are people working away determined to prove him wrong.

40
= 2 x 2 x 2 x 5
= 2 x 20
= 4 x 10
= 5 x 8
= 6 + 7 + 8 + 9 + 10

The word *forty* is the only number in English to have all of its letters arranged in alphabetical order.

40 is the fourth star number.

Forty winks means a short sleep.

A ruby anniversary celebrates 40 years.

Quadrille is a game played with 40 cards.

Quarantine is a period of isolation imposed on people and animals to prevent the spread of infectious disease. Cats, dogs and other animals coming into the UK must go into quarantine to prevent the spread of rabies unless they have a PETS certificate. Originally the word 'quarantine' meant a period of 40 days. Ships arriving in fourteenth century Venice were held for observation in case there were diseases on board. The choice of 40 days was probably made more for Biblical than for medical reasons.

According to the Bible, the rain fell for 40 days and 40 nights during Noah's flood. This has led to the English legend that if it rains on St Swithin's Day (15 July) there will be rain for 40 days. Needless to say, the changeable British climate does not support this prediction.

The Bible contains many references to events that lasted for 40 days and the expression just means a long period of time. Forty is frequently used as a convenient round number by the Semitic peoples, as in the tale of *Ali Baba and the Forty Thieves*.

Life begins at 40 is a catch phrase popular among those who in all honesty would rather be a little younger.

UB40 are a Birmingham based reggae group with UK hits between 1980 and 1988. The 1980s were a time of high unemployment in the UK and anyone seeking unemployment benefit had to fill out a government UB40 form which gave the band its name.

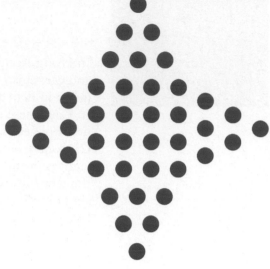

40 as a star number. The pattern is made up of a square and four equilateral triangles.

The Celsius and Fahrenheit temperature scales come together at minus forty. So -40°C is the same bitterly cold temperature as -40°F.

In Shakespeare's *A Midsummer Night's Dream* Puck says 'I'll put a girdle round the earth in forty minutes'. The earth has a diameter of about 12756 km (7926 miles). How fast would Puck have to travel to go all the way round in 40 minutes?

The Field of Forty Footsteps is an old name for some land at the back of the British Museum in London. The tradition is that two brothers fought each other there until both were killed and, for many years afterwards, 40 impressions of their feet remained in the field where no grass would grow. The fight was watched by the young woman who was the object of their contest and the bank where she sat also remained grassless.

There are 40 rods in one furlong. Rods and furlongs are old imperial units of distance that are only rarely used today. Rods are sometimes also called 'poles' or 'perches'. A furlong equals 201.17 metres and was originally a 'furrow long': the length of the furrow in the common field used in medieval farming.

What kinds of things could you have measured in rods? Is your height more or less than one rod?

41 is a prime number.

41 is the international telephone dialling code for Switzerland.

> Lizzie Borden took an axe
> And gave her mother 40 whacks;
> When she saw what she had done
> She gave her father 41.

Lizzie Andrew Borden (1860 – 1927) was popularly believed to have murdered her stepmother and her father, but was found not guilty at her trial.

This pattern is made from 41 tiles. It uses dark and light tiles with nine tiles across.

In the town square there is a pattern like this which is 19 tiles across. How many tiles will it contain altogether? Hint – try some smaller numbers first.

A pattern with 41 tiles.

Taking differences

There is a useful trick for investigating number sequences. For example, here is a sequence of square numbers –

169 196 225 256 289 324

What is the next number in the sequence?

The trick is to take the difference between each pair of numbers, and to go on doing this, until you get the same difference each time.

169 196 225 256 289 324
27 29 31 33 35
2   2   2   2

In the second row –
196 – 169 = 27,
225 – 196 = 29,
– and so on.

In the third row –
29 – 27 = 2,
31 – 29 = 2,
– and so on. In the third row, the difference is always two.

What is the next number in the original sequence of squares? It is just a matter of working backwards. The next number in the bottom row must be 2, so 37 is the next number in the middle row, and 361 must be the next number in the original sequence.

This trick is very powerful and can be used for lots of different problems. For some number sequences you have to take only one row of differences before you get a constant answer. For others, you will find that you need to take differences several times.

The trick does not always work and you may find you run out of numbers before you get a constant set of differences.

Other interesting things can happen as well. For example –

2   3   5   7   10   15   22   32
2   3   5   7   10

In this case, when you take differences, the sequence repeats itself.

You can work out how this sequence continues in a similar way to before. The next number in the lower row must be 15, and so the next number in the sequence must be 47.

Here are a few sequences you might like to investigate –

A: 5   8   11   14   17   20   23
B: 24   54   96   150 216 294 384
C: 4   8   16   32   64   128 256
D: 15   34   65   111 175 260 369
E: 7   11   18   29   47   76   123

(A is a linear sequence, B are the numbers of unit squares on the faces of cubes, C are the powers of two, D are the constants for magic squares of increasing size, and E are Lucas numbers.)

42
= 2 x 3 x 7
= 2 x 21
= 3 x 14
= 6 x 7
= 3 + 4 + 5 + 6 + 7 + 8 + 9
= 9 + 10 + 11 + 12
= 13 + 14 + 15
= 2 + 4 + 6 + 8 + 10 + 12

Forty-two is a domino game where the total value of the tricks is 42 points.

There are 42 holes in a *Connect 4* board.

■ According to Douglas Adams in *The Hitch Hiker's Guide to the Galaxy*, the number 42 is the answer 'To the great Question of Life, the Universe and Everything'.

Why 42? Douglas Adams and radio producer Geoffrey Perkins comment 'Many people have asked whether the choice of 42 as the Ultimate Answer came from Lewis Carroll or perhaps from an ancient Tibetan mystical cult where it is the symbol of truth. In fact it was simply chosen because it was a completely ordinary number, a number not just divisible by two but also by six and seven. In fact it's the sort of number that you could, without any fear, introduce to your parents.'

The Ultimate Answer is rumoured to have given its number to the group *Level 42*. Founded in 1980 this London based jazz/funk instrumental group fronted by Mark King had chart hits between 1983 and 1987.

■ *The Egyptian Book of the Dead* written around 2400 BC is the oldest religious book in existence. It is an account of life after death and the number 42 puts in an appearance.

'I know the names of the 42 gods who live with thee in the Hall of Maati.'

These 42 gods were said to be present at the last judgment of every human being.

■ *Forty-second Street* is a Hollywood musical which was made into a film in 1933. It tells the story of a Broadway musical producer who overcomes troubles during rehearsal with everything coming together on the opening night. Forty-second Street is a real street in the theatre district of New York.

■ Lewis Carroll had a special fondness for 42 and the number often occurs in his writing. In its first edition, *Alice's Adventures in Wonderland* contained 42 illustrations and included this delightful piece of nonsense -

> At this moment the King, who had been for some time busily writing in his notebook, called out "Silence!" and read out from his book "Rule Forty-two. All persons more than a mile high to leave the court." Everybody looked at Alice.
> "I'm not a mile high," said Alice.
> "You are," said the King.
> "Nearly two miles high," added the Queen.
> "Well, I shan't go, at any rate," said Alice: "besides, that's not a regular rule: you invented it just now."
> "It's the oldest rule in the book," said the King.
> "Then it ought to be Number One," said Alice.

In Lewis Carroll's *The Hunting of the Snark*, the baker 'had forty-two boxes all carefully packed, with his name painted clearly on each ... all left behind on the beach.'

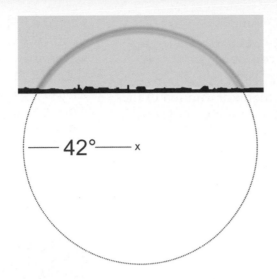

You can think of a rainbow as a circle in the sky, most of which is hidden below the horizon. The radius of this circle is an angle of 42° and it is directly opposite the sun. If your head could cast a shadow, it would lie in the centre of the circle. Rainbows are formed when sunlight is reflected off the inside of drops of water in the sky. The colours, ranging from red outside to violet inside, are caused because different wavelengths of light are reflected at slightly different angles.

■ 'A bicycle, certainly, but not the bicycle,' said he. 'I am familiar with 42 different impressions left by tyres.'

These words of the great fictional detective Sherlock Holmes would have been impressive in Victorian times, but 42 types of tyre would hardly be adequate today with the modern diversity of road transport.

■ Imagine a straight tunnel through the centre of the earth. Drop an object into the tunnel, and if there is no air resistance, it will emerge 42 minutes later on the other side of the world.

All 27 rows and columns of this magic cube add up to 42. The four main diagonals also each add up to 42.

There is one number which you cannot see because it is hidden right in the middle of the magic cube. Can you figure out what it is?

9	13	20
23	3	16
10	26	6

9	23	10	10	26	6	6	16	20
11	7	24	24	1	17	17	21	4
22	12	8	8	15	19	19	5	18

8	15	19
12	25	5
22	2	18

22	2	18
11	27	4
9	13	20

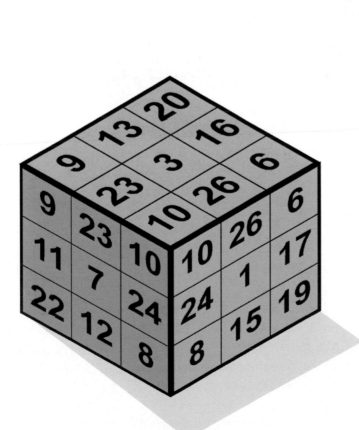

A magic cube with a constant of 42.

43 is a prime number.

> She may very well pass for forty-three
> In the dusk with the light behind her.

W. S. Gilbert (1836 - 1911) wrote this less-than-flattering description of the rich attorney's daughter in the comic opera *Trial By Jury*.

■ The youngest person elected to be President of the USA was John Fitzgerald Kennedy who took the oath of office aged 43 years in 1961. He was assassinated at Dallas in 1963. The youngest British Prime Minister was William Pitt who was aged 24 years when he took office in 1759. The youngest for over a century is Tony Blair who was also 43 when he became Prime Minister in 1997.

44
= 2 x 2 x 11
= 2 x 22
= 4 x 11
= 2 + 3 + 4 + 5 + 6 + 7 + 8 + 9
= 8 + 10 + 12 + 14

44 is the international telephone dialling code for the United Kingdom.

The number of different ways in which five people with five hats can each take the wrong hat is 44.

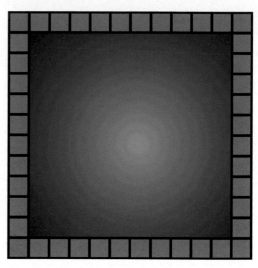

44 paving slabs around a pond.

A square pond 10 feet by 10 feet requires 44 paving slabs measuring one foot square to surround it.

How many paving slabs are needed for a pond measuring 5 feet by 5 feet? Can you find a rule that gives an answer for any size of pond?

This modern plate design arranges 44 flowers in a grid pattern.

45
= 3 x 3 x 5
= 3 x 15
= 5 x 9
= 1 + 2 + 3 + 4 + 5 + 6 + 7 + 8 + 9
A triangular number.

A *45* was a gramophone record that played at 45 revolutions per minute (or 45 rpm).

A sapphire anniversary celebrates 45 years.

45 is the international telephone dialling code for Denmark.

The Forty-Five is the name given to the rebellion of 1745 led by Charles Edward Stuart, the Young Pretender.

Forty-five is a card game played between two players or two teams. The winner is the first to gain 45 points.

This Mystic Rose pattern contains 45 lines. It was drawn by placing ten marks around a circle and joining all the marks together.

Somebody makes a bigger and better Mystic Rose by drawing lines between 20 positions around a circle. Without drawing their pattern, work out how many lines they would use.

Hint – try some smaller numbers first.

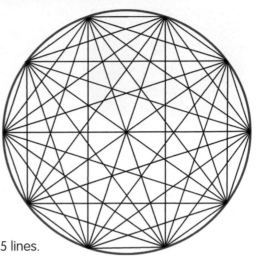

A Mystic Rose with 45 lines.

46
= 2 x 23
= 10 + 11 + 12 + 13
46 is the international telephone dialling code for Sweden.

Human beings have 46 chromosomes in every cell of their body. 23 of these are inherited from the mother and 23 from the father.

47 is a prime number.
1, 3, 4, 7, 11, 18, 29, 47...
A Lucas number.
47 is the international telephone dialling code for Norway.
Add 2 to 47 and you get 49. Multiply

2 by 47 and you get 94, which are the same digits reversed. The same works with larger numbers –
47 + 2 =   49   47 x 2 =   94
497 + 2 =   499   497 x 2 =   994
4997 + 2 = 4999   4997 x 2 = 9994
– and the pattern continues.

48
= 2 x 2 x 2 x 2 x 3
= 2 x 24
= 3 x 16
= 4 x 12
= 6 x 8

= 3 + 5 + 7 + 9 + 11 + 13
48 is the smallest number to have ten different factors. They are 1, 2, 3, 4, 6, 8, 12, 16, 24 and 48.
Suzi Quatro's *48 Crash* reached No 3 in the UK charts in 1973.
■ There are 48 douzièmes in one barleycorn.
Douzièmes and barleycorns are old imperial units of length which are no longer in use.
12 douzièmes make one line,
4 lines make one barleycorn,
3 barleycorns make one inch.
If 1 inch = 25.4mm, how long is a douzième? What sort of things could you measure with it?

■ *The 48 Preludes and Fugues* are 48 pieces of music written by Johann Sebastian Bach (1685 – 1750) which make use of all the major and minor musical keys. Bach was showing off the possibilities of a new discovery which he called the *Well-Tempered Clavier* and which we know as the modern piano keyboard.

With 48 matchsticks you can make a triangle in 48 different ways. Two of the possible triangles are shown below.

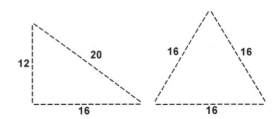

In how many ways could you make a triangle using 11 matches? And in how many ways using 12 matches?

49

= 7 x 7

A square number.

= 1 + 3 + 5 + 7 + 9 + 11 + 13

49 is a square number. Put 48 in the middle and you get 4,489 which is also square. Put 48 in the middle again and you get 444,889 which is another square number. Do it again and you find that 44,448,889 is also square.

Lasca is an unusual checkers game played on a seven by seven board with 49 squares.

49 is the international telephone dialling code for Germany.

One forty-ninth expressed as a decimal is

0.02 04 08 16 32 65 ...

After 42 digits the sequence starts to repeat.

Forty-niners were those who took part in the Californian gold rush of 1849 –

> In a cavern, in a canyon,
> Excavated for a mine,
> Dwelt a miner, forty-niner,
> And his daughter Clementine.

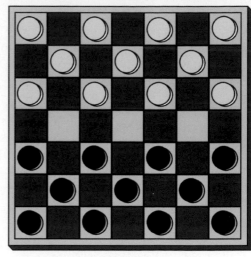

49 squares on a lasca board.

Forty-ninth Parallel was an award winning war film made in 1941 about five stranded German U-boat men trying to escape from Canada into the USA.

Bingo calls for 49 include 'Copper' (from PC 49, an old police radio series) and 'Cannock-nuff' (from the French quarante-neuf).

50

= 2 x 5 x 5

= 2 x 25

= 5 x 10

= 8 + 9 + 10 + 11 + 12

The Romans wrote 50 as L.

A golden anniversary celebrates 50 years.

Pentecost means 50th. Pentecost is a Jewish summer festival held on the 50th day after the Passover. It is often better known as Whitsun (or White Sunday).

50 metres freestyle is the shortest distance for a swimming event at the Olympic Games.

Fifty Ways to Leave Your Lover was a No 1 hit in the UK for Paul Simon in 1976.

To split something *fifty-fifty* means to divide it exactly in half.

In the United Kingdom the electricity supply has a frequency of 50Hz. This means that the voltage changes from positive to negative and back again 50 times a second.

In darts the inner bull scores 50.

> I asked my mother for fifty cents
> To see the elephant jump the fence;
> He jumped so high he reached the sky
> And didn't come back till the Fourth of July.

■ *Seven sevens are fifty?* was the title of an inaugural lecture by a professor of mathematics education called Hugh Burkhardt. It wasn't that the professor had forgotten his multiplication tables. He just wanted to make the point that in the real world it is sometimes more important to do an approximate calculation than an accurate one, and 7 x 7 = 50 is correct to one significant figure of accuracy.

50 Chinese lanterns make a pattern of 8 + 9 + 10 + 11 + 12.

■ The flag of the United States of America has 50 stars representing the 50 states. The flag has changed 28 times from the original version of 1775. After the declaration of independence in 1776 the flag had 13 stars and 13 stripes representing the 13 colonies which broke away from British rule. It was originally planned that each new member of the union should add one star and one stripe to the flag, but it soon became clear this was impractical. The original 13 stripes were retained with only a star being added for each new state. The current version of the flag gained its fiftieth star in 1960 when Hawaii joined the union.

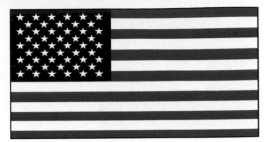

The stars in the flag can be counted in several different ways. If you count the diagonals you can see that
$1 + 3 + 5 + 7 + 9 + 9 + 7 + 5 + 3 + 1 = 50$

Imagine that a new state joined the USA and you were given the task of designing a new flag with 51 stars. How would you arrange the 51 stars to make an attractive pattern?

48 & 50 bottles

The Ilkeston Mineral Water Company was doing great business. They bottled water from a local spring and sold it to supermarkets and expensive restaurants at a great profit. But one day the owner, Mark Gradgrind, had a phone call from his biggest customer and he looked worried.

'They want us to stop sending them boxes with 48 bottles and start supplying them with boxes of 50 bottles. It's more convenient for them, they say, to have round numbers like 50. But what about our convenience? We have got thousands of those boxes. We will have to throw them all away.'

'Of course you won't have to throw them away' retorted Faye, who was Mr

Gradgrind's chief bottle washer.

'I've told you before you are packing them wrongly. You put them in straight rows. But if you used alternate rows of five and six bottles, the boxes would take two more bottles, and you would fit in fifty.'

Needless to say Mr Gradgrind took no notice, threw away all his old boxes, and still continued to make huge amounts of money.

But Faye was right. Can you see why? It works when you pack anything that is cylindrical like bottles and toilet rolls – you take up less space if you pack them on a hexagonal grid.

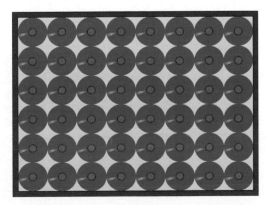

A box packed with 48 bottles arranged in 6 rows and 8 columns. Is it possible to fit in any more?

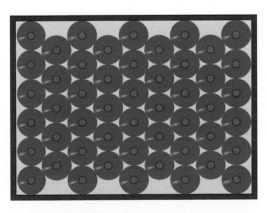

... 50 bottles packed into the same box by arranging them hexagonally.

51
= 3 x 17

= 6 + 7 + 8 + 9 + 10 + 11
= 1 + 4 + 7 + 10 + 13 + 16

To call someone a '51-ar' is to call them a liar. This slang expression comes from the roman numerals LI plus the letters AR.

52
= 2 x 2 x 13
= 2 x 26
= 4 x 13
= 3 + 4 + 5 + 6 + 7 + 8 + 9 + 10
= 10 + 12 + 14 + 16

There are 52 playing cards in a deck (without jokers).

There are 52 weeks in a year (and an odd day or two left over).

52 is the international telephone dialling code for Mexico.

There are 53 rings on this stand.

53 is a prime number.

If there are 53 people in a room it is very likely that at least two of them share the same birthday – that they were born on the same day and in the same month, but not necessarily in the same year. The slim chance that everybody has a different birthday is about 1 in 53.

One fifty-third expressed as a decimal has 13 digits that repeat –
0. 0188679245283 0188679245 ...

54
= 2 x 3 x 3 x 3
= 2 x 27
= 3 x 18
= 6 x 9
= 2 + 3 + 4 + 5 + 6 + 7 + 8 + 9 + 10

There are 54 playing cards in a deck (with jokers).

There are 54 panes of glass in these windows. You can see the pattern in different ways – as 6 x 9, or 3 x 18, or 2 x 3 x 9.

54 inches make one French ell. Cloth used to be measured in an old unit called an 'ell'. But people could not agree how long an ell was. In England it was 45 inches, in France it was 54 inches, and a Flemish ell was 27 inches.

What is the shortest length of cloth that would be an exact number of ells when measured in either the English, French or Flemish system?

54 small squares on a Rubik's cube.

55
$= 5 \times 11$
$= 1 + 2 + 3 + 4 + 6 + 7 + 8 + 9 + 10$
A triangular number.

1, 1, 2, 3, 5, 8, 13, 21, 34, 55 ...
A Fibonacci number.

$= 1 + 4 + 9 + 16 + 25$
A pyramidal number – the sum of the first five square numbers.

An emerald anniversary celebrates 55 years.

55 is the international telephone dialling code for Brazil.

£55,000 was the sum bet by Monsieur Phileas Fogg that he could go *Around the World in Eighty Days* in Jules Verne's novel published in 1873.

A set of 55 dominoes has extra tiles. The largest is 'double nine'. The normal set has just 28 tiles but you can buy sets with as many as 91 tiles.

A normal set of dominoes can be matched up to make a continuous chain that uses all the tiles. Why is this impossible with the set of 91?

Part of a set of 55 dominoes.

56
= 2 x 2 x 2 x 7
= 2 x 28
= 4 x 14
= 7 x 8
= 2 + 4 + 6 + 8 + 10 + 12 + 14
= 1 + 3 + 6 + 10 + 15 + 21
A tetrahedral number – the sum of the first six triangular numbers.

Part of a set of Tri-Ominoes.

Every one of the 56 triangular tiles in this set is different. They are used in a game called *Tri-Ominoes* where they have to be matched up like dominoes.

Someone buys a smaller set of these tiles which only uses the numbers between 0 and 3. How many tiles does the set contain?

■ The 56 Aubrey Holes are one of the oldest features of Stonehenge dating from about 2000 BC. They are a ring of 56 holes in a circle 284.5 feet in diameter placed with great accuracy. They must have been constructed for a clear purpose but today they are a mystery. One theory is that they once helped to predict when eclipses of the sun and moon would occur.

These old weights were used on a farm. One weighs 56 pounds and the other 28 pounds. Heavy objects were once weighed in hundredweights. One hundredweight equals 112 pounds. The 56 pound weight is half a hundredweight and the 28 pound weight is a quarter of a hundredweight.

57
= 3 x 19
= 7 + 8 + 9 + 10 + 11 + 12

57 Varieties is a trade mark for Heinz foods. It was devised in 1896 by the firm's founder, Henry Heinz, when travelling on a New York train. A card displayed in the train advertised a brand of shoes in '21 styles'. Heinz counted up his own products, and although they came to more than 57, the phrase *57 Varieties* appealed to him enough for it to stick.

In horse race betting a 'Heinz' is a combination of 57 bets on six horses. The bet is made up of 15 doubles, 20 trebles, 15 quadruples, six quintuples and one sextuple accumulator.

The Knights Templar travelling to fight in the crusades were required to recite the Lord's Prayer 57 times a day if they were unable to attend a church service.

There are 57 crossings on this Celtic plate.

58
= 2 x 29
= 13 + 14 + 15 + 16
= 10 + 13 + 16 + 19

The Welsh name for a town on Anglesey *Llanfairpwllgwyngyllgogerychwyrndrobwllllantysiliogogogoch* has 58 letters. It is the longest place name in the UK and means: St Mary's Church in the hollow of the white hazel near to the rapid whirlpool of Llantysilio of the Red Cave.

If a dentist gives you a gold filling it is in fact only 58% gold. The remainder is a mixture of silver and copper.

58° Celsius is the hottest air temperature ever recorded on earth. It was recorded in 1922 in Libya.

59 is a prime number.

Only 59% of the moon's surface can be observed from earth. Before the invention of space rockets, no one knew what the back of the moon looked like.

The planet mercury rotates once every 59 earth days.

One Bingo call for 59 is 'Brighton Line'. It has been suggested that the *Brighton Belle* train that travelled between London and Brighton was first pulled by locomotive number 59.

60
= 2 x 2 x 3 x 5
= 2 x 30
= 3 x 20
= 4 x 15
= 5 x 12
= 6 x 10

There are 60 seconds in one minute and 60 minutes in one hour.

Minutes and seconds are also used for measuring angles –
60 seconds of arc makes one minute, 60 minutes of arc makes one degree, 360 degrees makes one revolution.

The use of 60 (and 360) for measuring angles and time comes from Sumerian and Babylonian mathematics. More than 5000 years ago Sumeria was a thriving culture in the Middle East. They had a number system based on 60 which was very simple to use and made use of place value and fractions. 60 is a good choice for a number base because it can be divided equally in so many different ways. Its factors are 1, 2, 3, 4, 5, 6, 10, 12, 15, 20, 30 and 60. It is the smallest number to have 12 different factors.

The internal angles in an equilateral triangle measure 60°.

60 is the retirement age for women in the United Kingdom, although this is under review by the government.

60 mph is the UK national speed limit on single carriageways.

In old British money a crown was worth 60 pence.

A diamond anniversary celebrates 60 years.

SIXTY HORSES WEDGED IN A CHIMNEY

The journalist Beachcomber (J. B. Morton 1893 – 1979) invented this sensational newspaper headline and constantly regretted that the story to go with it had not turned up.

61 is a prime number.

One sixty-first expressed as a decimal is 0.01639344262295081967213 11... which continues as a repeating sequence of 60 digits.

61 is the international telephone dialling code for Australia.

61 is a hex number. This means you can arrange 61 coins into a hexagonal pattern with one coin in the middle.

61 as a hex number is shown in this pattern on the lid of a water butt.

. . . .

62
= 2 x 31
= 14 + 15 + 16 + 17

A distance of 100 kilometres is just a fraction over 62 miles

The 62 faces on this unusual solid are a mixture of decagons, hexagons and squares. Mathematicians call it a *truncated icosidodecahedron* but its friends just call it Sidney.

. . . .

63
= 3 x 3 x 7
= 3 x 21
= 7 x 9
= $2^0 + 2^1 + 2^2 + 2^3 + 2^4 + 2^5$
The sum of six powers of 2.

Queen Victoria reigned for 63 years from 1837 to 1901. She not only reigned longer than any other British monarch, but she is also the longest

---

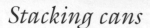

Stacking cans

Supermarkets stack up their cans in different ways. Surprisingly, both of the stacks in the pictures below contain exactly 27 cans. In the first stack, each layer of cans is a hexagon pattern.

Imagine a shop assistant builds taller stacks like these with one extra layer. For one stack, he begins with a 4 by 4 square of cans. For the other, he begins with a hexagon that has 4 cans on each side. When he has built them he is surprised to find they still have the same number. There are 64 cans in both stacks.

It turns out that the two types of stack have the same number of cans whatever height you build them. To understand why, it may help to look at the entry for the number 127.

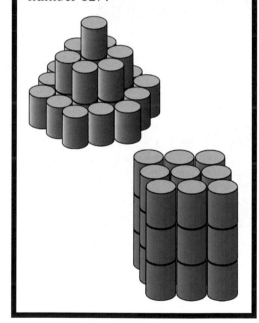

reigning queen anywhere in the world.

In Australia there are six coins in circulation: 5c, 10c, 20c, 50c, $1 and $2. If you had one of each of these you could make 63 different sums of money.

Two numbers have the same two digits in the opposite order. Their difference is 63. Can you find the numbers?

64
= 2 x 2 x 2 x 2 x 2 x 2
A sixth power.
= 4 x 4 x 4
A cube.
= 8 x 8
A square.
= 2 x 32
= 4 x 16
= 1 + 3 + 5 + 7 + 9 + 11 + 13 + 15
The first number after one to be both a square and a cube.

64 is the international telephone dialling code for New Zealand.

That's the sixty-four dollar question! is a catch phrase from radio and television quiz shows. It means 'I find that question very difficult to answer'. Originally it came from an American CBS radio quiz show called *Take It or Leave It* broadcast between 1941 and 1948 which had a top prize of $64. Since then a number of television quiz shows in Britain, America and elsewhere have adopted the phrase and inflation has transformed it into the more familiar *That's the sixty-four thousand dollar question!*

> Will you still need me,
> Will you still feed me,
> When I'm sixty-four?

This song by John Lennon and Paul McCartney is on The Beatles' *Sgt Pepper's Lonely Hearts Club Band* album. When they recorded the song in 1967, Lennon was aged 27, McCartney was 25, and an age of 64 must have seemed impossibly old. But age creeps up on everyone and McCartney should celebrate his 64th birthday in 2006.

There are 64 different codons. A codon is a unit of information in the genetic code which determines the characteristic form of every plant and animal. A codon is a group of three bases along a molecule of DNA or RNA. As there are four types of bases (T, A, G or C), there are 4 x 4 x 4 = 64 possible codons.

1 ♘	48	31	50	33	16	63	18
30	51	46	3	62	19	14	35
47	2	49	32	15	34	17	64
52	29	4	45	20	61	36	13
5	44	25	56	9	40	21	60
28	53	8	41	24	57	12	37
43	6	55	26	39	10	59	22
54	27	42	7	58	23	38	11

There are 64 squares on a chess board, 32 black and 32 white. Look at the chess board in the illustration above. If you follow the numbers from 1 to 64 you will move like a knight visiting every square on the board. This is called a 'knight's tour'. All the even numbers are on black squares and all the odd numbers on white squares. As well as this, the numbers form a kind of magic square. Every row and column adds up to 260.

65
= 5 x 13
= 2 + 3 + 4 + 5 + 6 + 7 + ... + 11

65 is the fifth star number.

The rows, columns and diagonals of 5 x 5 magic squares all add up to 65.

65 is the retirement age for men in the United Kingdom.

> I'm 65 and I guess that puts me in with the geriatrics. But if there were fifteen months in every year, I'd only be 48. That's the trouble with us. We number everything. Take women, for example. I think they deserve to have more than twelve years between the ages of 28 and 40.
> – James Thurber (1894 – 1961)

In the ASCII code used by computers, 65 represents a capital letter A.

Eiffel 65 is an Italian dance/pop group. Their debut single *Blue Da Ba Dee* was number one in many European countries in 1999.

65 is the international telephone dialling code for Singapore.

A party of students went for a Chinese meal. Every two people shared a dish of rice, every three people shared a dish of soup, and every four people shared a dish of chicken. If 65 dishes were eaten altogether, how many students were in the party?

· ·

66
= 2 x 3 x 11
= 2 x 33
= 3 x 22
= 6 x 11
= 1 + 2 + 3 + 4 + ... + 11

A triangular number.

66 is the international telephone dialling code for Thailand.

There are 66 books in the Bible – 39 in the Old Testament and 27 in the New Testament.

> If you ever plan to motor west,
> travel my way, take the highway that is best.
> Get your kicks on Route Sixty-Six.
>
> It winds from Chicago to LA,
> more than two thousand miles all the way.
> Get your kicks on Route Sixty-Six.

You Get Your Kicks on Route Sixty-Six was written by the American song writer Bobby Troup (1918 – 1999). This legendary road in the USA was opened in 1926 and today is replaced by modern interstate highways. In the 1930s John Steinbeck (1902 – 1968) also wrote about the road in *The Grapes of Wrath* –

> Highway 66 is the main migrant road. 66 – the long concrete path across the country, waving gently, up and down on the map, from the Mississippi to Bakersfield – over the red lands and the gray lands, twisting up into the mountains, crossing the Divide and down into the bright and terrible desert, and across the desert to the mountains again, and into the rich California valleys.

66 is the path of a people in flight, refugees from dust and shrinking land, from the thunder of tractors and shrinking ownership, from the desert's slow northward invasion, from the twisting winds that howl up out of Texas, from floods that bring no richness to the land and steal what little richness is there. From all of these the people are in flight, and they come into 66 from the tributary side roads, from the wagon tracks and the rutted country roads. 66 is the mother road, the road of flight.

This skeleton tower is six blocks high and has been built from 66 blocks. How many blocks would there be in a similar tower which was 12 blocks high?

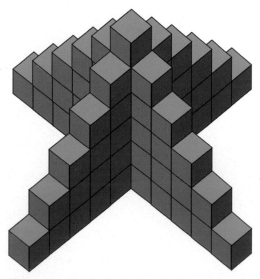

There are 66 blocks in this skeleton tower.

67 is a prime number.

67 is the number of boots worn by three football teams, one of which has a camel as a goal keeper. (The camel is always losing one of its boots.)

With just 12 cuts it is possible to divide a pizza into 67 tiny pieces. Most people would rather just eat it.

One sixty-seventh written as a decimal is 0.014925... and continues as a repeating sequence of 33 digits.

68
= 2 x 2 x 17
= 2 x 34
= 4 x 17
= 5 + 6 + 7 + 8 + 9 + 10 + 11 + 12
= 14 + 16 + 18 + 20

Paul Geidel served a record-breaking prison sentence of 68 years 8 months and 2 days for murder. He was released in 1980 in New York aged 85.

69
= 3 x 23
= 9 + 10 + 11 + 12 + 13 + 14

69 is the only number whose square and cube together use all the digits from 0 to 9 –

$69^2 = 4,761$
$69^3 = 328,509$

69 is the fifth number that stays the same when written upside down.

69 is the record for the most children born to one mother. It is held by a Russian peasant who lived in Shuya in the eighteenth century. In 27 pregnancies she gave birth to 16 pairs of twins, 7 sets of triplets and 4 sets of quadruplets.

'Sixty-nine!' – New Zealand sheep shearers use this phrase as a secret warning to each other to moderate their language when women or visitors are in ear shot.

70
= 2 x 5 x 7
= 2 x 35
= 5 x 14
= 7 x 10
= 7 + 8 + 9 + 10 + 11 + 12 + 13

Three score years and ten is traditionally regarded as the natural human life span. Today we think in terms of *life expectancy* which is the average number of years a newborn baby can expect to live. In Western Europe, North America and Australasia, life expectancy is about 77 years. In most African countries it is less than 55 years.

The national speed limit in the UK is 70 mph on dual carriageways and motorways.

A platinum anniversary celebrates 70 years.

71 is a prime number.

Last month there was a special offer on Taylor's crisps. Shops would give you a free packet of crisps in exchange for eight empty packets. Zoe Blackadder wasted no time and went into action right away. She scoured all the waste bins in town and collected 71 empty packets. How many free packets of crisps did Zoe collect?

Prime numbers

The first twelve prime numbers are –
2, 3, 5, 7, 11, 13, 17, 19, 23, 29, 31, 37 ...
A number is called prime if its only factors are one and itself.

From the entries in this book you will see that many numbers can be made by multiplying smaller numbers together. For example –

21 = 3 x 7

– 3 and 7 are called factors of 21. But some numbers cannot be made in this way and these are called prime. For example, 23 is a prime number because it cannot be made by multiplying together smaller numbers. Numbers like 21 which are not prime are sometimes called composite numbers.

All prime numbers, apart from 2, are odd numbers.

The Mersenne primes are a special type of prime number. The first five are –

3, 7, 31, 127, 8191

– and they all equal a power of two minus one –

$2^2 - 1 = 3$
$2^3 - 1 = 7$
$2^5 - 1 = 31$
$2^7 - 1 = 127$
$2^{13} - 1 = 8191$

For a mathematician, the equivalent of breaking the 100 metres world record is to find the highest known prime number. Every year or so, someone discovers a higher one and it gets reported in the newspapers. These record-breaking numbers are always Mersenne primes. At the time of writing this book the highest known prime is $2^{13466917} - 1$. To write it out you would use 4,053,946 digits and probably get through quite a few pencils.

Whenever someone discovers a new Mersenne prime they also automatically discover a new perfect number. $(2^{13466917} - 1) \times 2^{13466916}$ is currently the world's largest known perfect number.

It has been proved that the number of primes is limitless and so records for the highest known prime can go on being broken for ever.

In this number pyramid, all the primes from 2 to 619 are shown in black. The square numbers are shown in red.

The square numbers show a clear pattern but the prime numbers are much more disorganised. The columns are either odd or even, and 2 is the only prime number that appears in an even column. There are unbroken diagonals of primes running from 47 to 257 and from 223 to 593, but tidy patterns like this never continue for very long.

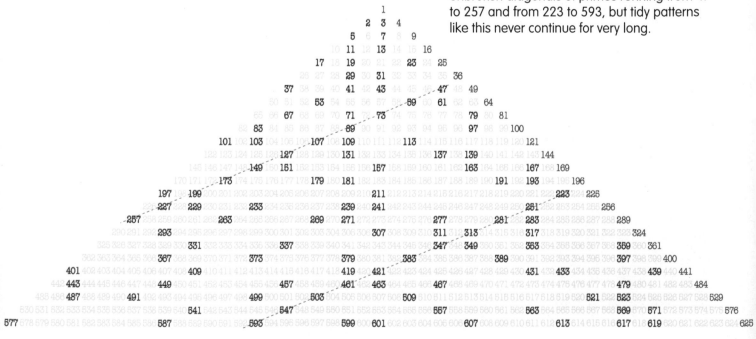

72
= 2 x 2 x 2 x 3 x 3
= 2 x 36
= 3 x 24
= 4 x 18
= 6 x 12
= 8 x 9
= 2 + 4 + 6 + 8 + 10 + 12 + 14 + 16
= 4 + 5 + 6 + 7 + 8 + 9 + 10 + 11 + 12

'Half a gross' means 72.

The normal human pulse rate is about 72 beats per minute at rest.

The Cambodian alphabet is the longest in the world with 72 letters.

The life expectancy for men in the United Kingdom is about 72 years.

The Bayeux Tapestry has 72 scenes. It tells the story in pictures of the Norman invasion of England in 1066.

How many different smells are there? According to the poet Samuel Taylor Coleridge (1772 – 1834) in *Cologne* there are at least 72 –

I counted two and seventy stenches,
All well defined, and several stinks!

72 is the only number which is eight times the sum of its own digits. What is the only number which is five times the sum of its own digits?

73 is a prime number.

In the past, 'The man on the Clapham omnibus' was someone who was typical of us all, who represented the concerns and opinions of ordinary people. He was a universal figure, an Everyman. But what about the bus? Without a doubt, the universal bus, the bus that captures the essence of all buses, is a number 73. It remains a mystery why a whole generation of comedians and writers chose 73 as the archetypal bus route. But there can be no doubt that, as 49 is to policemen, 73 is to buses.

In amateur radio, 73 is a shorthand for 'best wishes'. A radio ham might finish a conversation with 'OK, Tony, seven-three and I will talk to you later'. The convention may be an old telegraphic abbreviation from the days when messages were sent in morse code.

74
= 2 x 37

= 17 + 18 + 19 + 20

Tungsten, with the atomic number 74, is the metal in the filament of light bulbs. In a lamp it glows white-hot for hundreds of hours without breaking or melting. Tungsten melts at 3370°C which is the highest melting point of any metal.

75
= 3 x 5 x 5
= 3 x 25
= 5 x 15
= 3 + 4 + 5 + 6 + 7 + 8 + ... + 12

'75 per cent' means the same as 'three quarters'.

According to the Highway Code, the overall stopping distance for a car travelling at 30 mph is 75 feet.

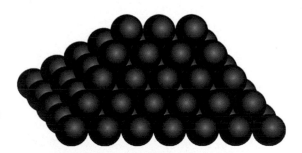

75 as a rectangular pyramid.
The base is a 7 x 5 rectangle,
the second layer is 6 x 4,
the third is 5 x 3,
the fourth 4 x 2,
and the top layer is a line of 3.

76

= 2 x 2 x 19

= 2 x 38

= 4 x 19

= 6 + 7 + 8 + 9 + 10 + 11 + 12 + 13

1, 3, 4, 7, 11, 18, 29 ... 76

A Lucas number.

Halley's comet appears every 76 years.

A bingo call for 76 is 'Trombones' from the song *Seventy-six trombones* which is featured in the American musical *The Music Man*.

To Americans '76' means 1776, the year of the Declaration of Independence. In the USA *76* is the name for a place in Kentucky.

77

= 7 x 11

Say 'seventy-seven' aloud and you use five syllables. It is the lowest number to use five syllables in English. The next lowest is 'one hundred and one'.

In a trivia quiz 9 points are awarded for a correct answer and 6 points are taken away for a wrong answer. Is it possible for someone to score 77? If not, how close can the score get to 77?

78

= 2 x 3 x 13

= 1 + 2 + 3 + 4 + 5 + 6 + ... + 12

A triangular number.

A *78* is an old type of gramophone record playing at 78 revolutions per minute.

The life expectancy for women in the United Kingdom is about 78 years.

■ This is the last verse from a well known Christmas song –

On the twelfth day of Christmas,
My true love sent to me
Twelve lords a-leaping,
Eleven ladies dancing,
Ten pipers piping,
Nine drummers drumming,
Eight maids a-milking,
Seven swans a-swimming,
Six geese a-laying,
Five gold rings,
Four colly birds,
Three French hens,
Two turtle doves, and
A partridge in a pear tree.

Adding these up, 78 presents were sent on the twelfth day. But how many presents were sent over all twelve days?

■ Tarot cards are probably the forerunners of modern playing cards. The full deck of 78 tarot cards is made up of 56 cards of the lesser arcana and 22 cards of the greater arcana. The cards are often used by fortune tellers.

The cards in the lesser arcana resemble modern playing cards. There are four suits called coins, swords, cups and batons (or clubs). In each suit there are cards numbered from 1 to 10 and four court cards: king, queen, knight and jack. The cards in the greater arcana have names: 1 The Magician, 2 The Lady Pope, 3 The Empress, 4 The Emperor, 5 The Pope, 6 The Lovers, 7 The Chariot, 8 Justice, 9 The Hermit, 10 The Wheel of Fortune, 11 Strength, 12 The Hanging Man, 13 Death, 14 Temperance, 15 The Devil, 16 The Tower (illustrated here), 17 The Star, 18 The Moon, 19 The Sun, 20 Judgement, 21 The World, and The Fool. The Fool is traditionally unnumbered, or sometimes numbered zero. He has become the Joker in modern playing cards.

Pascal's triangle

Pascal's Triangle is a pattern of numbers which is easy to make and is also very useful because it can be used to calculate all kinds of things. Every number is the sum of the two numbers above it and the ends of the rows are always '1'.

The triangle is named after the French mathematician Blaise Pascal (1623 – 1662) who described the pattern although he did not discover it. The earliest known reference to it appears in a 1303 Chinese book *The Precious Mirror of the Four Elements* by Chu Shih-chieh.

```
                        1
                      1   1
                    1   2   1
                  1   3   3   1
                1   4   6   4   1
              1   5  10  10   5   1
            1   6  15  20  15   6   1
          1   7  21  35  35  21   7   1
        1   8  28  56  70  56  28   8   1
      1   9  36  84 126 126  84  36   9   1
    1  10  45 120 210 252 210 120  45  10   1
  1  11  55 165 330 462 462 330 165  55  11   1
1  12  66 220 495 792 924 792 495 220  66  12   1
```

> **1 + 6 = 7**
>
> Each number in the triangle is made by adding together the two numbers above it.

Choose one of these liquorice allsorts. You have seven to choose from.

But suppose you can choose two of them. How many choices do you have now? Choosing the two coconut ones would be one choice. Choosing the two striped ones would be a different choice. Or you could take the two pink ones – that would be another choice. There are in fact 21 different ways of choosing two out of seven liquorice allsorts.

If you could choose three out of seven the number of choices goes up to 35. With four out of seven there are also 35 choices, and with five out of seven you have 21 ways to choose. With six out of seven there are only seven ways to choose, as you really only have to decide which one to leave behind.

These numbers, 7, 21, 35, 35, 21, 7, are one of the rows in Pascal's Triangle. The triangle can be used to calculate the number of *combinations* of things, like the number of different ways of choosing liquorice allsorts.

For example, if you are the manager of a netball club, and you have to choose a team of seven from ten possible players, in how many ways can you do this? Look for the row in Pascal's Triangle that begins 1, 10.... and count across seven times starting from the '10'. You will find there are 120 different ways the team can be chosen.

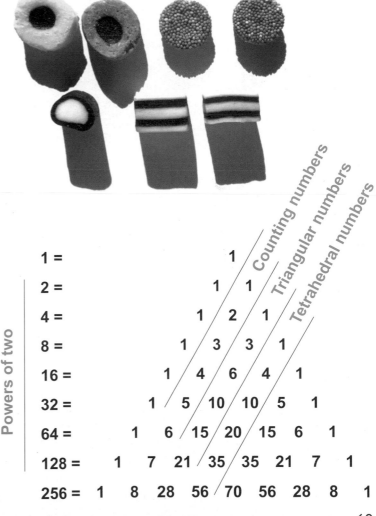

Powers of two

1 =						1						
2 =					1		1					
4 =				1		2		1				
8 =			1		3		3		1			
16 =		1		4		6		4		1		
32 =		1	5		10		10		5		1	
64 =	1		6	15		20		15	6		1	
128 =	1	7	21		35		35	21	7		1	
256 =	1	8	28	56		70	56	28	8		1	

Counting numbers · Triangular numbers · Tetrahedral numbers

79 is a prime number.

79 = (7 + 9) + (7 x 9)

All two digit numbers ending in nine are equal to the sum of their digits added to the product of their digits.

The yellow metal Gold has the atomic number 79. This means that there are exactly 79 protons in each atom of gold. The metal is best known for making jewellery but it is also used in dentistry and the manufacture of electronic components.

80

= 2 x 2 x 2 x 2 x 5

= 2 x 40

= 4 x 20

= 5 x 16

= 8 x 10

= 3 + 5 + 7 + 9 + 11 + 13 + 15 + 17

80 chains make a mile.

An octogenarian is someone aged between 80 and 89 years.

Tea bags are often sold in boxes of 80. Other common sizes are 40 and 160.

Around the World in Eighty Days is a book by Jules Verne, and also a television series by Michael Palin made in 1989.

In the early days of personal computers, the number 80 had a strange fascination for computer manufacturers. It was apparently unthinkable to produce a computer that did not have 80 in its name... TRS80, ZX80, 380Z.

The French for 80 is *quatre-vingts* which translates as 'four twenties'. This seems strange to us because we are used to number systems that count in multiples of ten, but the French language is not alone in sometimes counting in twenties. The Mayans and the Aztecs made extensive use of multiples of twenty in their counting systems and, in Europe, a number of Celtic languages including Scottish Gaelic also do this. Modern Irish counts in this way –

20	fiche	(twenty)
40	da fiche	(two twenties)
60	tri fiche	(three twenties)
80	ceithre fiche	(four twenties)
...		
160	ocht fiche	(eight twenties)
180	naoi fiche	(nine twenties)

Fashions for counting come and go. Before the eleventh century, the French counted, as we do, in tens, but by the seventeenth century multiples of 20 were used all the way from 60 *(trois-vingts)* up to 360 *(dix-huit vingts)*. In Paris there is an old hospital built by Louis IX which is still called the *Quinze-vingts* after its 300 blind inmates. Today some local dialects of French still retain phrases like *huit-vingts* for 160, but *quatre-vingts* is all that remains in formal French.

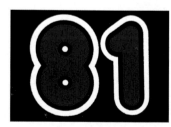

81

= 3 x 3 x 3 x 3

A fourth power.

= 3 x 27

= 9 x 9

A square number.

= 1 + 3 + 5 + 7 + 9 + 11 + ... + 17

81 is the international telephone dialling code for Japan.

> A dozen, a gross, and a score,
> plus three times the square root of four,
> divided by seven,
> plus five times eleven,
> equals nine squared and not a bit more.
> — Jon Saxton

One eighty-first as a decimal is –
0.012345679 012345679 012345679 ...
which continues for ever. It is curious that the digit eight is missing.

There are 81 small squares on this Canadian road sign which means 'road closed'.

Powers of three:
the first figure has three circles,
the second has nine,
the third has 27
and the fourth has 81.

The Hinton Bicycle Company prided itself on the brightly-coloured wheels it put on its bicycles. For the front wheel you could choose between cerise, orange, turquoise, lime-green, indigo, gamboge, terracotta, electric-blue and violet. The back wheel could also be in any of the same nine colours. As you were allowed to order a bicycle with any combination of back and front wheels, this gave you a choice of 81 different models.

After trading for several years, the Manageress of the Hinton Bicycle Company realised that no one ever ordered a cerise front wheel, and also that no one ever wanted a turquoise back wheel. So it was decided that these two options would be discontinued.

After this change, how many different models were available?

Powers

The 'fifth power of 3' simply means the number you get when you multiply 3 together five times. So it is –
3 x 3 x 3 x 3 x 3
– which equals 243.

This can also be written 3^5 where the little '5' means 'to the fifth power'.

The 'first power' of a number is always equal to itself. So the first power of 3 equals 3. The 'second power' means the same as the square of a number. So the second power of 3 equals 9. The 'third power' is another way of saying the cube of a number.

Here are some of the powers of three –
$3^1 = 3$
$3^2 = 3 \times 3 = 9$
$3^3 = 3 \times 3 \times 3 = 27$
$3^4 = 3 \times 3 \times 3 \times 3 = 81$
$3^5 = 3 \times 3 \times 3 \times 3 \times 3 = 243$

The 'zeroth power' of any number always equals one, so $3^0 = 1$.

The easiest powers to work out are the powers of ten –
$10^0 = 1$
$10^1 = 10$
$10^2 = 100$
$10^3 = 1,000$
$10^4 = 10,000$
$10^5 = 100,000$
$10^6 = 1,000,000$
$10^7 = 10,000,000$
...

It is easy to see how this pattern continues.

Powers of ten have a special use for writing very big numbers. For example, the speed at which light travels is about 300,000,000 metres/second. But it is much easier to write –
3×10^8 metres/second
– 10^8 tells us the number of zeros after the 3.

82

= 2 x 41

= 19 + 20 + 21 + 22

There are 82 different ways of linking six hexagons together. A figure made from six hexagons is called a hexahex.

■ The metal lead has the atomic number 82. It was one of the earliest metals to be discovered because it is easy to extract and to work.

The Latin word for lead is *plumbum* from which we get the chemical symbol Pb and the word plumber, someone who traditionally works with lead.

Lead is more common than most other heavy chemical elements. This is because it lies in an 'island of stability' in the table of elements, with a nucleus of 82 protons that is less likely to be broken up than its neighbouring elements. The next 'island of stability' occurs at atomic number 114.

There is increasing concern about the risk of poisoning from lead compounds in the air and in drinking water. Lead is known to be a poison which accumulates in the body over a long period.

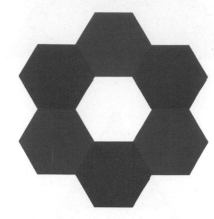

This is just one of 82 possible ways of joining together six hexagons.

83 is a prime number.

The French writer Voltaire, who was born in 1694, lived to be 83. This was a good age for someone living at a time when the average life expectancy was about 35 years.

Q – How many times can you subtract 7 from 83, and what is left afterwards?
A – You can subtract it as many times as you want, and it leaves 76 every time.

84

= 2 x 2 x 3 x 7

= 2 x 42

= 3 x 28

= 4 x 21

= 6 x 14

= 7 x 12

= 1 + 3 + 6 + 10 + 15 + 21 + 28

A tetrahedral number – the sum of the first seven triangular numbers.

A pregnant woman has a 1 in 84 chance of having twins.

84 Charing Cross Road is a book by Helene Hanff which has been turned into a play and a film. It is the story of a New York woman who conducts a long correspondence with a London antiquarian bookseller.

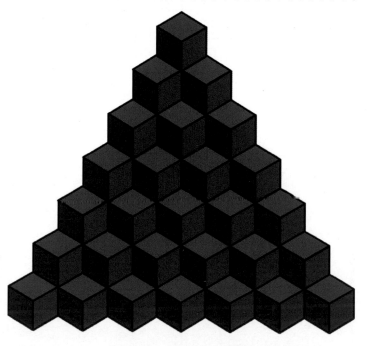

You can count the number 84 in two different ways in this figure. You would need 84 diamond-shaped tiles to make the pattern, and as a 3-dimensional structure, you could build it with 84 blocks.

85

= 5 x 17

= 4 + 5 + 6 + 7 + 8 + 9 + ... + 13

The 85 Ways to Tie a Tie is a book by Thomas Fink and Yong Mao. It brings together the history of tie knots with a branch of mathematics called knot theory. Although the book does indeed describe 85 ways of making a tie knot, the authors claim that only 13 of these are stylish.

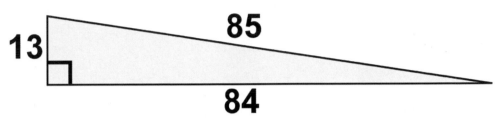

■ This is a right-angled triangle with sides of 13, 84 and 85. It does not matter whether these measurements are inches or metres – it will still be right-angled because of Pythagoras's theorem –

$$13^2 + 84^2 = 85^2$$

– the sum of the squares on the two shorter sides equals the square on the longest side.

Draw a circle inside this right-angled triangle so that it touches all three sides (an *incircle*). Its diameter equals the sum of the two shorter sides minus the longest side –

13 + 84 - 85 = 12

Groups of numbers like 13, 84 and 85 are called 'Pythagorean triples' because they can make the sides of a right-angled triangle. The best known Pythagorean triple is 3, 4 and 5.

There is a neat pattern to some of the Pythagorean triples. These figures give the size of right angled triangles and the diameter of their incircles –

$3^2 + 4^2 = 5^2$	3 + 4 - 5 = 2
$5^2 + 12^2 = 13^2$	5 + 12 - 13 = 4
$7^2 + 24^2 = 25^2$	7 + 24 - 25 = 6
$9^2 + 40^2 = 41^2$	9 + 40 - 41 = 8
$11^2 + 60^2 = 61^2$	11 + 60 - 61 = 10
$13^2 + 84^2 = 85^2$	13 + 84 - 85 = 12

– and the pattern continues.

Can you predict the next line?

86

= 2 x 43

= 20 + 21 + 22 + 23

86 is the international telephone dialling code for China.

Eighty-six is a slang expression used by workers in restaurants and bars, particularly in America. It can mean that the restaurant is out of something ('we're eighty-six on soup') or, that

something should be got rid of ('eighty-six that cat – the health inspector is outside'). When a customer is 'eighty-sixed' he is refused service. There are many theories about the origin of the expression. The *Oxford English Dictionary* suggests *eighty-six* may have been rhyming slang for *nix*, meaning nothing.

87

= 3 x 29

= 12 + 13 + 14 + 15 + 16 + 17

By tradition, 87 is an unlucky score for cricketers. This is perhaps because 87 is 13 runs short of a century. Unlucky is a relative term, however, and there are some cricket teams who are grateful if they ever reach a score of 87.

Diamonds are printed 87 times on a normal deck of playing cards. There are also 87 of the symbols for the other suits. An ace has three symbols printed on it, a two has four symbols, and so on, up to ten which has twelve symbols. The court cards each have four symbols.

88

= 2 x 2 x 2 x 11

= 2 x 44

= 4 x 22

= 8 x 11

= 3 + 4 + 5 + 6 + 7 + 8 + 9 + 10 + 11 + 12 + 13

There are 88 keys on a modern piano, 52 white and 36 black, spanning more than seven octaves. The piano has a greater musical range than any other orchestral instrument.

The Japanese have a special name for the very large number that is written as one followed by 88 zeros (10^{88}). They call it *muryoutaisuu* which translates as 'a large amount of nothing'.

88 is the 6th number that stays the same when written upside down.

Pianos used to have 85 keys spanning seven octaves (from A to A) but modern pianos have three extra keys making a total of 88 keys.

89 is a prime number.

1, 1, 2, 3, 5, 8, 13, 21, 34, 55, 89 ...
A Fibonacci number.

One eighty-ninth as a decimal is –
0.0 1 1 2 3 5 9 5 ...
– curiously the sequence begins with
the Fibonacci numbers, but this
pattern breaks down at '9'. There is a
recurring sequence of 44 digits.

■ The seedhead of a sunflower usually
has 55 spirals going clockwise and 89
spirals going anticlockwise. It is no
coincidence that both 55 and 89 are
Fibonacci numbers. Spirals of this kind
are quite common in plants and nearly
always occur in numbers from the
Fibonacci sequence.

Smaller sunflowers have only 34 and
55 spirals. There have been reports of
bigger ones with 89 and 144 spirals.

The pattern itself does not change for
different sizes of sunflower. But as it
grows larger, our eyes tend to be
drawn to different spiral patterns
according to the size. You can see this
by looking at the photograph through
a hole in a piece of paper. If the hole is
cut so that only the centre of seedhead
is visible you will now see only 34 and
55 spirals as on a smaller sunflower.

The same pattern can be seen in pine
cones, daisies, pineapples and is also
present, but less prominent, in many
other plants. The pattern has evolved
to give an even packing of seeds as the
flower grows. The angle between one
seed and the next seed to emerge is
222.49° (or 360° divided by the
Golden Ratio 1.618).

89 and 55 spirals on a sunflower.

90
= 2 x 3 x 3 x 5
= 2 x 45
= 3 x 30
= 5 x 18
= 6 x 15
= 9 x 10

= 2 + 4 + 6 + 8 + 10 + ... + 18

A nonagenarian is a person in their nineties – aged between 90 and 99.

Bamboo, the fastest growing plant, can increase its height by 90 cm in one day.

Cassettes marked '90' have a playing time of an hour and a half.

'Ninety per cent of everything is rubbish' is a catchphrase coined by the cynical Theodore Sturgeon.

'The first 90 minutes are the most important.' This advice is often heard at the start of a football match which, of course, last 90 minutes.

> We hope to amuse the customers with music and with rhyme
> But ninety minutes is a long, long time.

Like many theatre people, Noel Coward (1899 – 1973) was lukewarm about the arrival of television. The lines come from his opening song for a 90 minute CBS television show made in 1955.

In Worcestershire dialect a *ninety bird* is someone of bad character. *Nineted* means notorious. "'E's a nineted un, 'e is."

■ 90 degrees make a right angle. Right angles are everywhere. You will find them at the corners of this book, in the corners of the room you are in, and in most of the furniture that surrounds you. Although designers sometimes try to break away from the right angle and produce objects based on other angles such as 60°, it is not easy to abandon the right angle completely, particularly if you want table tops that are horizontal and walls that are vertical.

Most right angles that we see are man made but some occur naturally. If you look at crystals of table salt under a microscope you will see they are perfect cubes, with right angles at each corner.

When we say something is 'square' we mean that its angles are very close to 90°. Our eyes seem especially sensitive to small deviations from 90° and we are easily irritated by shelves that are slightly crooked or fences that are askew.

A range of instruments have been devised to help us produce things with accurate right angles. The ancient Egyptians had an ingenious method for getting a square corner to a field. Knots were made at equal intervals along a rope and it was joined to make a loop with 12 sections. When the rope was pegged out on the ground it made a right-angled triangle with sides of length 3, 4 and 5 units.

This trick works because 3, 4 and 5 meet the conditions of Pythagoras's theorem – that a triangle has a right angle when the square of the longest side of the triangle equals the sum of the squares of the other two sides. Because –
$$3^2 + 4^2 = 5^2$$
– the triangle must have a right angle.

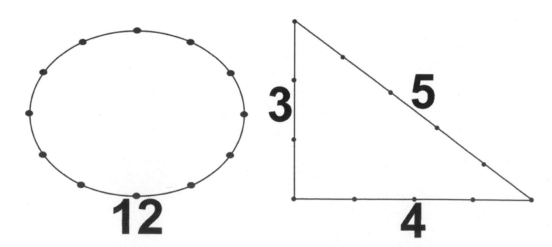

Pegged out a knotted rope to make a right-angled triangle.

91

= 7 x 13

= 1 + 2 + 3 + 4 + 5 + 6 + ... + 13
A triangular number.

1 + 4 + 9 + 16 + 25 + 36
A pyramidal number – the sum of the first six square numbers.

91 is the international telephone dialling code for India.

Queen's Guard is an old game played on a board in the shape of a hexagon made up of 91 small hexagons in alternating light and dark bands.

A special set of 91 dominoes has extra tiles. The largest is 'double twelve'. A normal set of dominoes has just 28 tiles.

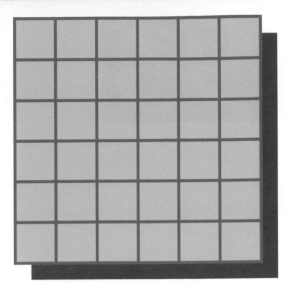

How many squares can you see here? There are 36 small squares, but there are also some larger ones including the whole pattern itself. Adding them all up, the total comes to 91 squares.

Draw a 4 x 4 square and count up all the squares in the same way. How many?

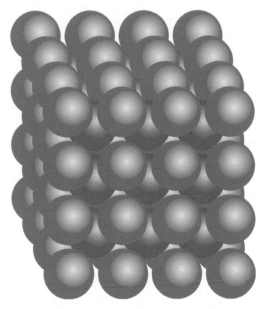

91 is a centred cube number. It is the sum of two consecutive cubes.
$4^3 + 3^3 = 64 + 27 = 91$.

92

= 2 x 2 x 23

= 2 x 46

= 4 x 23

= 8 + 9 + 10 + 11 + ... + 15

92 is the international telephone dialling code for Pakistan.

In Thailand you will find 92 different species of bats.

There are ninety-two naturally occurring chemical elements. Number 92 is the metal uranium with the atomic number 92.

93
= 3 x 31
= 13 + 14 + 15 + 16 + 17 + 18
In this pattern there are 93 circles based on a 17 x 17 square.

How many circles with a 7 x 7 square? Can you find a rule that gives the number of circles for any size of square?

94
= 2 x 47
= 22 + 23 + 24 + 25
94 is the international telephone dialling code for Sri Lanka.

The Irish playwright George Bernard Shaw, who was born in 1856, lived to be 94 years old.

Plutonium, with the atomic number 94, is a notorious chemical element which does not exist naturally and was first created artificially in 1940. It is used to make atomic weapons and in nuclear reactors. It is very poisonous and is radioactive. It decays very slowly with a half life of 24,000 years. This means that 4 kg of plutonium will decay to 2 kg after 24,000 years and to 1 kg after 48,000 years.

95
= 5 x 19
Martin Luther's *95 Theses* which gave birth to the Reformation were nailed to the door of the church in Wittenberg in 1517.

96
= 2 x 2 x 2 x 2 x 2 x 3
= 2 x 48
= 3 x 32
= 4 x 24
= 6 x 16
= 8 x 12
= 5 + 7 + 9 + 11 + 13 + 15 + 17 + 19

96 is the seventh number that stays the same when written upside down.

96 is the sixth star number.

If you are arrested by the police in England and Wales, you can be held for up to 96 hours (four days) without being charged.

Ninety-six is the name of a place in South Carolina, USA.

Courier Chess was played on a board with 96 squares. It was a variation of chess first recorded in thirteenth century Germany. The board had 8 rows and 12 columns. Both players had 24 chessmen including a king, a queen, a sage, a jester, two couriers, two bishops, two knights, two rooks and twelve pawns. Each piece had its own moves. The bishop moved two squares diagonally and could jump over other pieces.

Jamilla has four skirts, eight tops and three pairs of shoes. As these all go together, she figures out that she has 96 different outfits. She buys another pair of shoes which she can wear with any of her other clothes. How many complete outfits does she have now?

96 squares on a garden trellis.

97 is a prime number.

One ninety-seventh written as a decimal is –
0. 01 03 09 27 83 ...
The first four pairs of digits are the powers of three but that pattern does not carry on. If you continue long enough you find a repeating sequence of 96 digits.

'Ninety-seven horse power' is the refrain from a song by Michael Flanders (1922 – 1975) and Donald Swann (1923 – 1994) in praise of a London bus –

That monarch of the road,
Observer of the Highway Code,
That big six-wheeler
Scarlet-painted
Diesel-engined
Ninety-seven horse power
Omnibus!

98
= 2 x 7 x 7
= 2 x 49
= 7 x 14
= 11 + 12 + 13 + 14 + 15 + 16 + 17
= $1^4 + 2^4 + 3^4$

The sum of the first three fourth powers.

One ninety-eighth written as a decimal begins with the powers of two –
0. 01 02 04 08 16 32 65 30 ...
– but the pattern breaks down with 65. There is a repeating sequence of 42 digits.

The city of Hong Kong has the highest population density in the world with about 98 thousand people per square kilometre.

The normal body temperature is about 98° Fahrenheit.

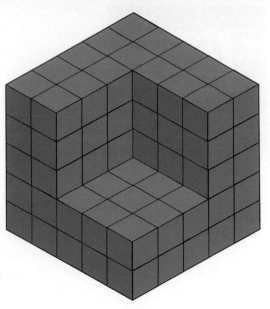

98 is the difference between two cubes. It is 5^3 - 3^3. What other numbers are the difference between two cubes?

99
= 3 x 3 x 11
= 3 x 33
= 9 x 11
= 4 + 5 + 6 + 7 + 8 + 9 + 10 + ... + 14
99^2 = 9801 and 98 + 01 = 99

Q – What goes '99 bonk'?
A – A centipede with a wooden leg.

99 Red Balloons was a hit for the German group Nena in 1984 and reached No 1 in the UK charts. They released the same song in the USA and Australia in a German version *99 Luftballoons*. This reached No 2 in the US singles chart.

In Islam there are 99 Most Beautiful Names Of God and there are also 99 noble names of the prophet Muhammad.

■ 'Say 99' used to be a doctor's instruction to a patient when listening to the chest with a stethoscope. The nasal sound of the words 'ninety nine' make a resonant sound that was supposed to help in diagnosis. What would a French doctor have asked a patient to say? The direct translation 'quatre-vingts dix-neuf' is such a mouthful that it might risk doing the patient an injury. In fact, the French equivalent is 'trente-trois' which has a similar nasal resonance.

■ Write down any three digit number (e.g. 841). Reverse its digits and take the difference (841 - 148 = 693). The answer will always be a multiple of 99 (7 x 99 = 693). Can you explain why?

■ A *99* is an ice cream cone with a chocolate flake. Cadbury's have been making flake chocolate bars since the 1920s. In 2002 Cadbury sold over 99 million 99 flake bars. On a hot summer's day about 6 flakes are sold every second. The origin of the name is uncertain but it is thought that in the early days Cadbury's did not give the product a name because it only had trade sales. 99 flake was originally just a product code because its recipe was recorded in ledger number 99.

Try working these out on a calculator –
9 x 9 = ?
99 x 99 = ?
999 x 999 = ?
9999 x 9999 = ?

Can you guess how the pattern will continue?

100
= 2 x 2 x 5 x 5
= 2 x 50
= 4 x 25
= 5 x 20
= 10 x 10
A square number.
= $1^3 + 2^3 + 3^3 + 4^3$
The sum of the first four cubes.

Q – What lies on its back, one hundred feet in the air?

A – A dead centipede.

There are several species of centipede but few have as many as 100 legs. Exceptionally there is one species of centipede which can have as many as 177 pairs of legs.

The Romans wrote a hundred as C. This is an abbreviation for the Latin word *centum*, and from this come many words meaning a hundred or a hundredth.

In Europe, a cent is a hundredth of a euro and, in North America, it is a hundredth of a dollar. One per cent means one hundredth.

A centenarian is someone who is a hundred years old and a century is a hundred years or a hundred runs at cricket.

100° is the boiling point of water on the Celsius scale. This used to be called the centigrade scale, but it has been renamed in honour of its inventor the Swedish astronomer Anders Celsius (1701 – 1744).

With measurements *centi-* always means one hundredth so 100 centimetres make a metre and 100 centiseconds make a second. *Centi-* is abbreviated to 'c', e.g. cm means centimetre.

Hecto- means a hundred times so another name for 100 metres is a hectometre. *Hecto-* is sometimes abbreviated to 'h', e.g. hV means hectovolt.

A *ton* means 100 pounds in money or a speed of 100 mph.

'Let 100 flowers bloom and 100 schools of thought contend' was the command given by the Chinese leader Mao Tse Tung in 1956. For a brief period he encouraged an open debate about the future of communism in China, but soon thought better of it, and free speech was repressed once again.

■ Athletes in the 100 metres race reach speeds in excess of 20 mph (32 km/h). When Florence Griffith-Joyner broke the women's 100 metres world record in 1988, she recorded a peak speed of 24.58 mph (39.56 km/h). Surprisingly the average speed over 200 metres can be faster than over 100 metres because of the time it takes to respond to the starting pistol and build up speed. In 2002 the men's 100 metres world record was 9.78 seconds (Tim Montgomery), but the 200 metres record was 19.32 seconds (Michael Johnson), which is a faster average speed.

The world record for 100 metres on a unicycle is 12.11 seconds made by Peter Rosendahl of Sweden in 1994.

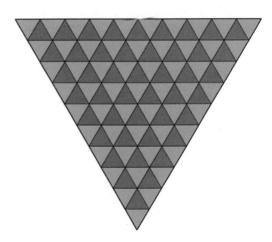

A pattern that is built from 100 small triangles. Any square number can be made into a pattern like this.

Spaghetti numbers, pizza numbers & potato numbers

Imagine you take a knife and make three straight cuts through a potato. After the third cut you take the pieces apart. How many pieces are there? The answer depends on where you make the cuts, but the greatest number of pieces you can produce is eight.

Now imagine making three straight cuts across a pizza. You try making the cuts in different places to get as many pieces as possible, but you are not allowed to move the pieces until you have made the final cut. This time the greatest number of pieces you can produce is only seven.

Next think of a straight piece of spaghetti. It is not difficult to see that three cuts will always produce four pieces of spaghetti.

There is nothing special about spaghetti, pizzas and potatoes, but they are good examples of 1-dimensional, 2-dimensional and 3-dimensional objects.

Potatoes and other 3-dimensional objects like blocks of wood, snow balls and cheeses can be divided into two pieces with one cut, four pieces with two cuts, eight pieces with three cuts. With each extra cut, the numbers of pieces are –
2, 4, 8, 15, 26 ...
We will call these 'potato numbers'.

2-dimensional objects like pizzas, pancakes and sheets of paper produce these numbers when you slice through them –
2, 4, 7, 11, 16 ...
We will call these 'pizza numbers'.

With spaghetti and other 1-dimensional objects like worms and pieces of string, you simply get one extra piece for each cut you make –
2, 3, 4, 5, 6 ...
We will call these 'spaghetti numbers'.

The triangle of numbers provides an easy way to work out potato, pizza and spaghetti numbers.

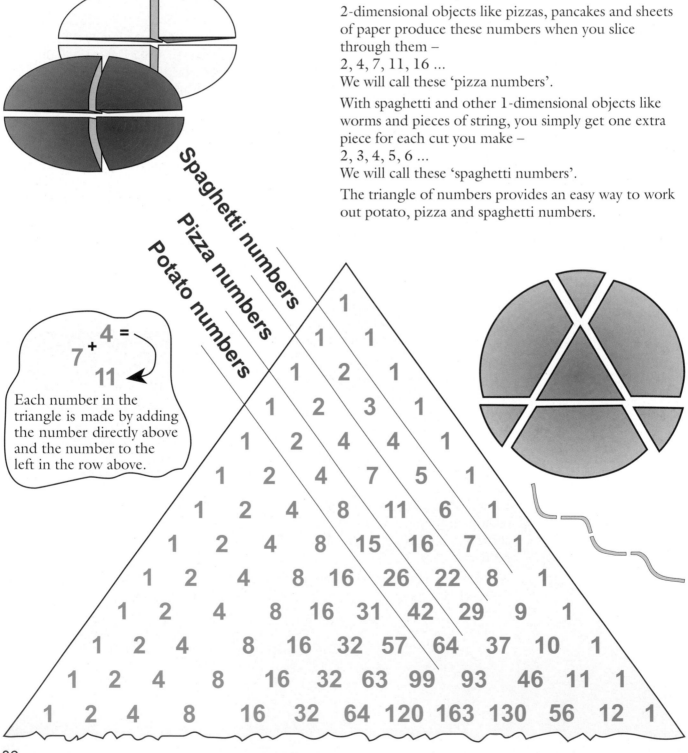

$$7 \overset{+}{} 4 = 11$$

Each number in the triangle is made by adding the number directly above and the number to the left in the row above.

82

101 is a prime number.

One Hundred and One Dalmatians (1912) is a novel by Dodie Smith. The story tells how the dogs of London help save puppies which are being stolen for their skins by the villainess Cruella De Vil. The book has been made into one of Walt Disney's best cartoon films (1961) and more recently a feature film with Glenn Close (1996). She also starred in a sequel *102 Dalmatians* (2000).

■ *Room 101* was the horrific invention of George Orwell (1903 – 1950) in his novel *1984* –

"The thing that is in Room 101 is the worst thing in the world."

Shortly before writing the book, Orwell worked for the BBC in an office with the room number 101 and it has been suggested that he took the name from this. The BBC has since had its revenge by using the name *Room 101* for a television comedy programme where celebrities name their deeply-felt hates. When he appeared on the show, Peter Cook condemned pet rabbits, Gracie Fields and the countryside to the flames of Room 101.

102
= 2 x 3 x 17
= 2 x 51
= 3 x 34
= 6 x 17

102 is the name of a river in Missouri, USA. To French explorers the Amerindian name for the river sounded like *cent deux*, the French words for 102.

103 is a prime number.

103 is the number of arms on thirteen octopuses one of whom worked as a careless guillotine operator.

104
= 2 x 2 x 2 x 13
= 2 x 52
= 4 x 26
= 8 x 13
= 6 + 8 + 10 + 12 + 14 + 16 + 18 + 20

A total eclipse of the moon lasts 104 minutes at most.

105

$= 3 \times 5 \times 7$

$= 3 \times 35$

$= 5 \times 21$

$= 7 \times 15$

$= 1 + 2 + 3 + 4 + 5 + 6 + 7 + ... + 14$

A triangular number.

A guinea is worth 105 pence, or in old British money, one pound and one shilling. Originally guineas were coins made from gold from Guinea in West Africa. They were used in England from 1663 until 1817. By tradition, lawyers and race horses are paid for in guineas rather than in pounds.

The Thousand Guineas is a famous horse race held at Newmarket. In 1814, when the first race was run, the original prize money was 1000 guineas or £1050.

In the UK it is well known that the Queen sends you a telegram when you celebrate your one hundredth birthday. It is less well known that the Queen also sends you a telegram when your birthday is 105, 106, 107 or more. Diamond wedding anniversaries also merit a telegram.

A launderette attendant collects the odd socks that get left behind in the washing machines. He has 15 socks, all in different colours and patterns, and none making a matching pair. He calculates that if he wears a different pair of odd socks every day, it would take him 105 days to work through all the possibilities. He has only just figured this out when he finds a sixteenth odd sock. How many extra days does he gain?

- -

106

$= 2 \times 53$

$= 25 + 26 + 27 + 28$

In theory, 15 straight cuts through a pizza can produce as many as 106 pieces.

- -

107 is a prime number.

107 appears in the sequence of numbers –

2, 7, 9, 16, 25, 41, 66, 107...

Each number is made by adding

together the two previous numbers, in a similar way to Fibonacci numbers. The sequence includes the three consecutive square numbers 9, 16 and 25 (3^2, 4^2 and 5^2).

- -

108

$= 2 \times 2 \times 3 \times 3 \times 3$

$= 2 \times 54$

$= 3 \times 36$

$= 4 \times 27$

$= 6 \times 18$

$= 9 \times 12$

$= 8 + 9 + 10 + 11 + 12 + ... + 16$

The interior angles in a regular pentagon measure 108°.

There are 108 different types of heptomino. Heptominoes are shapes made by fitting together seven squares. The illustration shows one of these

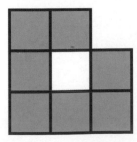

heptominoes which has a hole in the middle. It is the smallest polyomino to contain a hole.

The number 108 appears in Hindu and Buddhist writing. The god Krishna was said to have danced and played with 108 gopis or cowgirls. Buddhist prayer beads are arranged in groups of 108.

Strings of 108 beads have been used in several different cultures. These are court beads used by the emperors of the Qing Dynasty in China. Four groups of 26 lazurite beads are separated by single beads of a different size and colour. Additional beads hang loosely from the main string.

In the film *Bull Durham* (1988) Susan Sarandon plays the character Annie Savoy who makes a comparison between religion and the game of baseball – "There are 108 beads in a Catholic rosary and there are 108 stitches in a baseball." Coincidentally the film has a running time of 108 minutes.

You make 108 by multiplying together one one, two twos and three threes. It is one of a sequence of numbers that grows very rapidly –

1 x 2 x 2 = 4
1 x 2 x 2 x 3 x 3 x 3 = 108
1 x 2 x 2 x 3 x 3 x 3 x 4 x 4 x 4 x 4 = 27,648
1 x 2 x 2 x 3 x 3 x 3 x 4 x 4 x 4 x 4 x 5 x 5 x 5 x 5 x 5 = 86,400,000

109 is a prime number.

■ In 1873 the Belgian scientist Joseph Plateau made a startling discovery about soap bubbles. Although the bubbles in your bath tub may look disorganised, Plateau found that bubbles can only join each other at one of two angles. They always intersect at 109° or 120°. No other angles are possible.

The illustration shows how six soap bubble films can meet at the centre of a pyramid made from wire. There is an angle of 109° between each pair of lines which meet in the middle. Also each of these lines is formed by three bubble films coming together at angles of 120°.

Soap bubbles have a fascination for mathematicians. One of the leading workers in this field is Jean E. Taylor. She is one of a new generation of experimental mathematicians who make extensive use of computers and practical materials to do their mathematics.

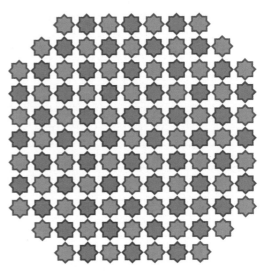

An arrangement of 109 stars.

■ Using only the coins in circulation in the UK, you can pay the sum of 30 pence in 109 different ways.

Paying with a 20p coin and a 10p coin would be one way. Giving thirty 1p coins would be another. There are 109 ways altogether using only 1p, 2p, 5p, 10p and 20p coins.

Here is the same information for other amounts of money, using the 50p and £1 coins as well.

You can give 1p in 1 way,
you can give 2p in 2 ways,
you can give 5p in 4 ways,
you can give 10p in 11 ways,
you can give 20p in 41 ways,
you can give 30p in 109 ways,
you can give 40p in 236 ways,
you can give 50p in 451 ways,
you can give £1 in 4563 ways.

In Australia they use 5c, 10c, 20c, 50c, $1 and $2 coins. In the USA they use 1c, 5c, 10c, 25c and 50c coins. In how many ways can you pay 20 cents in Australia? In how many ways can you pay 20 cents in the USA?

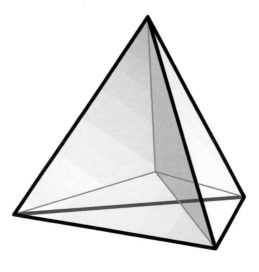

Soap bubbles meet at the centre of a pyramid made of wire.

110
= 2 x 5 x 11
= 2 x 55
= 5 x 22
= 10 x 11
= 2 + 4 + 6 + 8 + 10 + 12 + ... + 20

In the Olympic Games, the shortest men's hurdle race is raced over 110 metres. The shortest women's race is raced over 100 metres..

Computer keyboards have approximately 110 keys.

110 is a film size used in some popular miniature cameras.

This litter bin has a pattern of 110 holes.

A pattern with 110 circles.

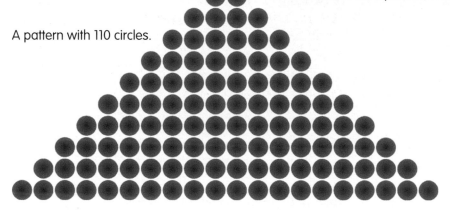

111
= 3 x 37
= 16 + 17 + 18 + 19 + 20 + 21

In cricket, a score of 111 is called 'the Nelson' and is considered unlucky. A 'double Nelson' is 222.

111 is the ninth number that stays the same when written upside down.

111 is the constant of a 6 x 6 magic square. This has the numbers from 1 to 36 arranged so that every row, column and diagonal adds up to 111.

Numbers like 111 which are just written with ones are called repunits. What happens when you square repunits? Try working these out on a calculator –

1 x 1 = ?
11 x 11 = ?
111 x 111 = ?
1111 x 1111 = ?

How does the pattern continue?

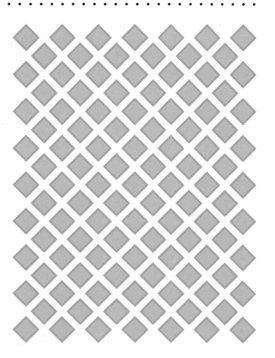

A pattern with 111 diamonds.

112
= 2 x 2 x 2 x 2 x 7
= 2 x 56
= 4 x 28
= 7 x 16
= 8 x 14

There are 112 lb in one hundredweight (cwt). Coal used to be sold in hundredweight sacks which you had to be strong to lift. It took 20 of these sacks to make a ton of coal. A hundredweight was a convenient measure because it was about the heaviest weight you would expect a man to lift. A hundredweight is equal to 50.80 kg. The modern equivalent is the metric hundredweight which is exactly 50 kg. In the USA a 'short hundredweight' is 100 lb.

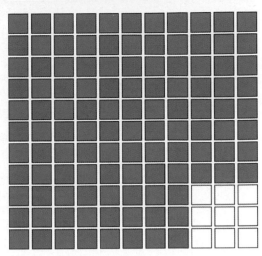

112 as the difference between two squares: $11^2 - 3^2$. What other numbers can be made from the difference of two squares?

This large chocolate bar is divided into 112 squares. Several people are going to share it. Can you divide it equally between 4 people? Between 5 people? Between 6 people? Or between 7 people?

113 is a prime number.

So are 131 and 311. Whichever way you arrange the digits of 113 you get a prime number. Not many numbers have this property. The only other three-digit numbers are 199 and 337, and their rearrangements.

Which two-digit prime numbers are also prime when you reverse their digits?

114
= 2 x 3 x 19
= 2 x 57
= 3 x 38
= 6 x 19
= 4 + 5 + 6 + 7 + 8 + 9 + 10 + ... + 15

The Muslim holy book the Koran is divided into 114 suras or chapters.

Ununquadium is the strange name given to chemical element number 114 which was first synthesised in 1998. It is one of the more stable artificially produced elements with a half life of about 30 seconds. At the time of writing it is the newest and heaviest element.

Very few people live to celebrate their 114th birthday. Two proven cases in the UK are Mrs Anna Williams (born 1873) and Miss Charlotte Hughes (born 1877). The authenticated world record is Jeanne Louise Calment of Arles in France who lived to an age of 122 and died in 1997.

If you celebrate your 114th birthday today, in which year were you born?

115
= 5 x 23
= 19 + 21 + 23 + 25 + 27

116
= 2 x 2 x 29
= 2 x 58
= 4 x 29
= 11 + 12 + 13 + 14 + ... + 18

117
= 3 x 3 x 13
= 3 x 39
= 9 x 13
= 1 + 4 + 7 + 10 + 13 + ... + 25

117 is a pentagonal number. The pentagonal numbers are 1, 5, 12, 22, 35, 51, 70, 92, 117, 145, 176 ...

Pentagonal numbers are a logical extension of the idea of triangular numbers and square numbers. The differences between the triangular numbers are 1, 2, 3, 4, 5, 6 ... The differences between the square numbers are the odd numbers 1, 3, 5, 7, 9, 11 ... And as you might expect the differences between the pentagonal numbers go up in threes 1, 4, 7, 10, 13, 16 ...

All pentagonal numbers are exactly one third of a triangular number. So for example 3 x 117 = 351, which is triangular. To demonstrate this, can you rearrange three copies of the shape shown in the diagram to make an equilateral triangle?

This diagram with 117 circles shows one way of representing pentagonal numbers. You can also see the smaller pentagonal numbers within this pattern. You may be wondering why there is no pentagon drawn here. The answer is that, although they are called pentagonal, it does not always help to think of a pentagon when visualising pentagonal numbers.

118
= 2 x 59
= 28 + 29 + 30 + 31

In the UK, the telephone prefix for directory enquiries.

How many triangles are here? There are 49 small triangles, but there are also several larger triangles, as well as the big triangle made by the pattern itself. If you add them all up the total comes to 118 triangles.

Draw a smaller version of this pattern with only four rows. How many triangles can you find in this? How does the number go up as you add on more rows?

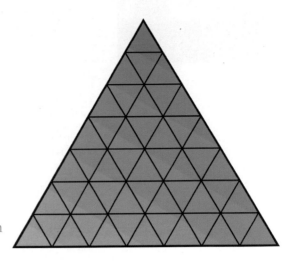

119
= 7 x 17
= 14 + 15 + 16 + 17 + 18 + 19 + 20

Factorials

When mathematicians write 5! the exclamation mark does not express surprise or excitement. It is just a convenient way of writing 'factorial 5'. This means the number you get by multiplying together –
1 x 2 x 3 x 4 x 5
– which comes to 120.

Factorials of other numbers are calculated in the same way. Factorial 10 or 10! is –
1 x 2 x 3 x 4 x 5 x 6 x 7 x 8 x 9 x 10
– which comes to quite a large number – 3,628,800.

Factorials are useful in working out the number of different ways you can arrange things. Factorial *n* is the number of ways *n* things can be arranged in a line. So if you had five garden gnomes in your front garden, they could be lined up in 5! or 120 different ways.

In the same way, eleven players in a football team can stand in a row in 11! ways and a deck of 52 playing cards can be ordered in 52! different ways which is a huge number with 68 digits.

Here are the first ten factorials –
1! = 1
2! = 2
3! = 6
4! = 24
5! = 120
6! = 720
7! = 5,040
8! = 40,320
9! = 362,880
10! = 3,628,800

120

= 2 x 2 x 2 x 3 x 5

= 2 x 60

= 3 x 40

= 4 x 30

= 5 x 24

= 6 x 20

= 8 x 15

= 10 x 12

= 1 x 2 x 3 x 4 x 5

Factorial 5 or 5!

= 1 + 2 + 3 + 4 + ... + 15

A triangular number.

= 1 + 3 + 6 + 10 + 15 + 21 + 28 + 36

A tetrahedral number – the sum of the first eight triangular numbers.

The interior angles in a regular hexagon measure 120°.

120 is the smallest number to have 16 different factors. These are 1, 2, 3, 4, 5, 6, 8, 10, 12, 15, 20, 24, 30, 40, 60, and 120.

Think of any five consecutive numbers (e.g. 752, 753, 754, 755, 756) and multiply them together. The answer will always be divisible by 120.

A cassette tape marked *120* has a playing time of two hours or 120 minutes.

120 is a film size used largely by professional photographers. Wedding photographs are often taken on this size of film.

You have five packets of crisps in different flavours. In what order are you going to eat them? You have 120 orders to choose from.

In old British money, *an angel* was a coin worth 120 pence.

■ Snooker uses a triangle of red balls. In a space ship where there is no gravity you could play a similar game with the balls arranged in a pyramid shape called a tetrahedron. But when the spaceship returned to Earth, gravity would make it necessary to go back to playing the two-dimensional game that uses a triangle of balls. If you were designing a game like this you might choose to use 120 balls because this number can be arranged exactly into either a triangle or a tetrahedron.

There is a number smaller than 120 that can also be arranged in these two ways. Can you find it?

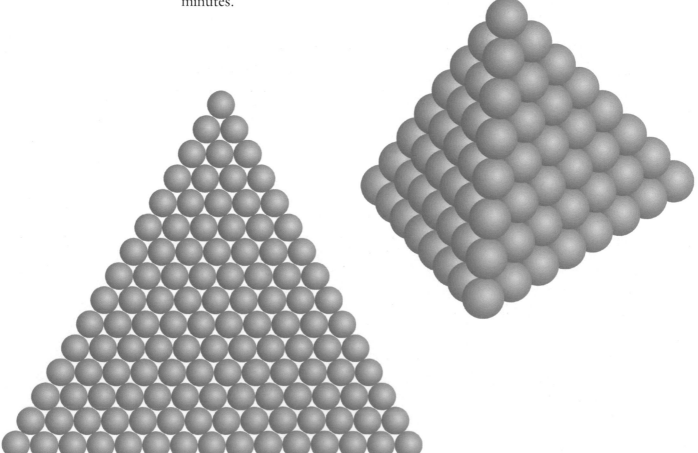

Galactic snooker – 120 balls arranged in a triangle and a tetrahedron.

■ 120 is sometimes called a *great hundred*. Fishermen in Norway and Germany used to measure their catch in great hundreds, counting 120 fish. In Iceland the word *hundrath* once meant 120 and not 100, until the arrival of Christianity introduced the modern usage and some confusion as both meanings continued to be used together for some years.

A market gardener displays a hundred of asparagus which contains 120 buds of the vegetable. The English Vale of Evesham has kept the tradition of the great hundred when counting asparagus.

121
= 11 x 11
A square number.
= 1 + 3 + 5 + 7 + 9 + 11 + 13 + ... + 21
= $3^0 + 3^1 + 3^2 + 3^3 + 3^4$
= 1 + 3 + 9 + 27 + 81
121 is a palindromic number because it reads the same backwards and forwards. It is also the square of another palindromic number (11).

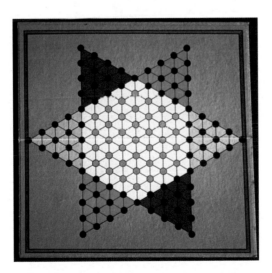

A Chinese Chequers board with 121 circles.

■ *Hex* is a game which uses a diamond-shaped board made from 121 hexagons. One player uses black counters and the other white, which they play in turn. The four edges of the board are also coloured black and white. The game is won by the first player to make a continuous line of counters joining their two sides of the board.

The game was invented in the 1940s by the Danish poet and engineer Piet Hein. Although you can buy Hex boards, it is easy to play with two coloured pens and a photocopy of the board.

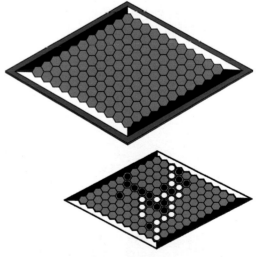

Boards for playing Hex.
In the smaller illustration, white has won by making a continuous line joining the two white edges of the board.

122
= 2 x 61
= 29 + 30 + 31 + 32

123
= 3 x 41
= 18 + 19 + 20 + 21 + 22 + 23
1, 3, 4, 7, 11, 18, 29... 123
A Lucas number.
Work these out with a calculator –
1 x 8 + 1 = ?
12 x 8 + 2 = ?
123 x 8 + 3 = ?
1234 x 8 + 4 = ?
12345 x 8 + 5 = ?
Can you explain the pattern that you get?

. . . .

124
= 2 x 2 x 31
= 2 x 62
= 4 x 31
= 12 + 13 + 14 + 15 + ... + 19

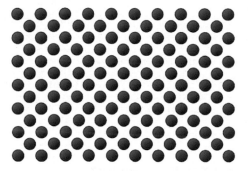

A pattern with 124 circles. This arrangement works because (7 x 10) + (6 x 9) = 124.

Lucas numbers

The first ten Lucas numbers are –
1, 3, 4, 7, 11, 18, 29, 47, 76, 123
The sequence is made in the same way as for Fibonacci numbers. You get each number by adding together the two previous numbers –
1 + 3 = 4
3 + 4 = 7
4 + 7 = 11
– and so on.

If you take pairs of Lucas numbers and divide the higher by the lower you get almost the same number each time –
18/11 = 1.6363..
29/18 = 1.6111..
47/29 = 1.6206..
76/47 = 1.6170..
123/76 = 1.6184..
– the answer gets closer and closer to a number called the Golden Ratio which equals 1.6180339887. This is also true for Fibonacci numbers.

. . . .

125
= 5 x 5 x 5
A cube.
= 5 x 25

When they were introduced in 1978, Intercity 125 trains were the pride of Britain's rail network. They were designed to run at 125 mph. In a test run in 1987, one of the trains reached a speed of 148 mph making it the fastest diesel train of its time. But this British technology was soon overtaken by France and Japan. In 1990 the French Train à Grande Vitesse (TGV) achieved a speed of 320 mph travelling between Courtalain and Tours.

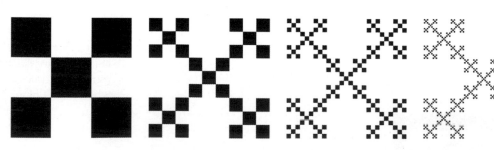

Going from left to right, each pattern has five times as many coloured squares as the previous pattern. The numbers of squares are: 1, 5, 25, 125 and 625, which are the powers of five.

126
= 2 x 3 x 3 x 7
= 2 x 63
= 3 x 42
= 6 x 21
= 7 x 18
= 9 x 14

126 is a film size used in easy-to-use cameras.

A vegan empties a barrel containing 126 litres of porridge using a 2-litre measure, a 3-litre measure and a 5-litre measure, filling each of these full each time it is used. She notices that the total amount she removed with the 3-litre measure was five times as much as she drew with the 2-litre measure. How much porridge did she remove with each measure?

Cubes

The first six cube numbers are –
1, 8, 27, 64, 125, 216
They are equal to –
1 x 1 x 1,
2 x 2 x 2,
3 x 3 x 3,
4 x 4 x 4,
5 x 5 x 5,
6 x 6 x 6
– which can be written more neatly as –
$1^3, 2^3, 3^3, 4^3, 5^3, 6^3$
– the small '3' means 'cubed'.

The cube of 5 is 125, and working backwards, we say the cube root of 125 is 5.

There is a connection between cube numbers and cube shapes. If you use cube shaped bricks to build a larger cube, the number of bricks you need is a cube number. For example, if you wanted to build an 8 cm cube using 1 cm cube bricks you would need 512 bricks which is the cube of 8.

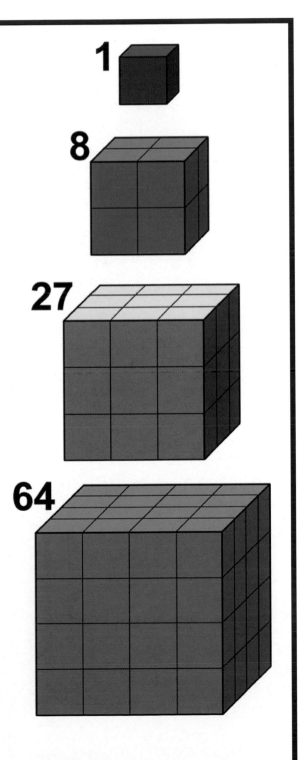

1 =	1	2	3	4	5	6
8 =	2	4	6	8	10	12
27 =	3	6	9	12	15	18
64 =	4	8	12	16	20	24
125 =	5	10	15	20	25	30
216 =	6	12	18	24	30	36

Cube numbers can be found in the multiplication table. Each L-shaped section has numbers that add up to a cube.

127

127 is a prime number.

$= 2^0 + 2^1 + 2^2 + 2^3 + 2^4 + 2^5 + 2^6$

The sum of seven powers of 2.

$= 2^7 - 1$

A Mersenne prime number.

If you have one each of the 1p, 2p, 5p, 10p, 20p, 50p and £1 coins you can make 127 different sums of money.

127 is the seventh hex number.

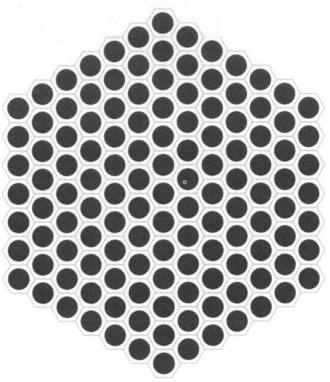

How many circles are there in this hex pattern with a side length of 7? There is a trick that saves the need to count them. With a little imagination you can see the pattern as a cube which also has a side length of 7. Concealed behind it will be a smaller cube with a side length of 6. Imagine you throw this away and you are left with the hex pattern. So the number of circles in the hex pattern is the difference between two cubes –
$7^3 - 6^3 = 343 - 216 = 127$.

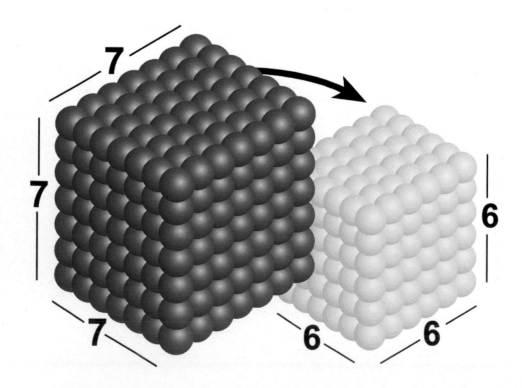

128
= 2 x 2 x 2 x 2 x 2 x 2 x 2
= 2^7

A seventh power.

1, 2, 4, 8, 16, 32, 64, 128 ...
The number made by starting with one and doubling seven times.

Your ancestors are likely to include 128 great-great-great-great-great-grandparents.

Acorn Master 128 computers were popular in UK schools in the 1980s. It had just 128 K of RAM memory. The number 128 often crops up with computers because they use binary arithmetic and 128 is a nice round number in binary notation – 10000000.

■ Many books have 128 pages and so does this one. Check out any books you have around and you will find that the number of pages is nearly always a multiple of 16. Besides 128, common numbers are 144, 160, 192 and 256 pages. Books are made by folding and cutting large sheets of paper. If you fold a sheet of paper three times you will get 8 leaves or 16 pages.

Carry out your own survey to find common numbers of pages in paperback books. Don't count the covers but be sure to count all the other pages, including any blank pages at the end. It will help you if the pages are numbered, but check for any unnumbered pages at the beginning and the end of the book as these should be included in your total. What is the commonest number of pages?

Powers of two

Powers of two are the doubling numbers –
2, 4, 8, 16, 32 ...
They can be calculated as –
2, 2 x 2, 2 x 2 x 2, 2 x 2 x 2 x 2,
2 x 2 x 2 x 2 x 2 ...
– which can be written more simply as –
2^1, 2^2, 2^3, 2^4, 2^5...

If you fold a sheet of paper in half and fold it again and again, each fold will double the number of thicknesses of paper. One fold produces two layers, two folds make four layers, and so on, going up in powers of two.

There is a famous legend about a Persian king who wanted to reward the man who invented the game of chess. When asked what he wanted, the man pointed to the chessboard. He asked for a single grain of wheat on the first square of the board, two grains of wheat on the second square, four grains on the third square, and so on in powers of two, up to the 64th square. This sounds quite a reasonable request, but if the king had agreed to it, he would have needed 9,223,372,036,854,775,808 grains for the 64th square, which is more than the world's annual wheat harvest today.

129
= 3 x 43
= 19 + 20 + 21 + 22 + 23 + 24

130
= 2 x 5 x 13
= 2 x 65
= 5 x 26
= 10 x 13

Ten straight cuts through a potato can produce as many as 130 pieces.

131 is a prime number.

How much earth is there in a hole in the ground which is 131 metres long, 131 metres wide and 131 metres deep?

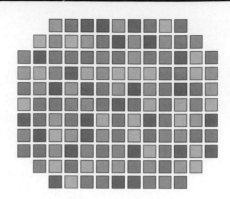

131 as a pattern of squares.

132
= 2 x 2 x 3 x 11
= 2 x 66
= 3 x 44
= 4 x 33
= 6 x 22
= 12 x 11

From the digits of 132 you can make 6 two digit numbers – 12, 13, 21, 23, 31 and 32. If you add these up you get 132.

133
= 7 x 19
= 16 + 17 + 18 + 19 + 20 + 21 + 22

133 is the seventh star number.

E133 on a food container means it contains a synthetic coal dye called *Brilliant Blue*. You may find it in some tins of processed peas. There are several hundred of these E numbers recognised by the EU to describe food additives. In the 1980s almost every food container had a list of its contents shown in this way, but it is less common now because food manufacturers are realising that the public really want to know what they are eating. Among the more unusual food additives listed as E numbers are pure gold (E175) and sulphuric acid (E513).

134
= 2 x 67
= 32 + 33 + 34 + 35

135
= 3 x 3 x 3 x 5
= 3 x 45
= 5 x 27
= 9 x 15
= 9 + 10 + 11 + 12 + 13 + 14 + 15 + 16 + 17 + 18
= $1^1 + 3^2 + 5^3$

The internal angles in a regular octagon measure 135°.

136
= 2 x 2 x 2 x 17
= 2 x 68
= 4 x 34
= 8 x 17
= 1 + 2 + 3 + 4 + ... + 16
A triangular number.

There is a curious link between the numbers 136 and 244 –
$1^3 + 3^3 + 6^3 = 244$
$2^3 + 4^3 + 4^3 = 136$

If 17 arm-wrestlers test their strength by challenging each other to a contest, there would be 136 arm-wrestling matches altogether.

137 is a prime number.

In theory, it is possible to divide a pizza into 137 tiny pieces by making just 17 straight cuts through it.

In Bertrand Russell's short story *The Mathematician's Nightmare*, the number 137 is an unruly rebel unwilling to accept its place in the number sequence.

138
= 2 x 3 x 23
= 2 x 69
= 3 x 46
= 6 x 23

139 is a prime number.

139 is a happy number. To find out whether a number is happy or not, you must square its digits, add them up, and go on doing this. If you eventually get to one, you have a happy number.

With 139 it takes five steps to reach one –
$1^2 + 3^2 + 9^2 = 91$
$9^2 + 1^2 = 82$
$8^2 + 2^2 = 68$
$6^2 + 8^2 = 100$
$1^2 + 0^2 + 0^2 = 1$
– and so 139 is happy.

What other happy numbers can you find? What happens with unhappy numbers? Hint – try some small numbers to begin with.

Triangular numbers

The first ten triangular numbers are –

1, 3, 6, 10, 15, 21, 28, 36, 45, 55

You can calculate triangular numbers by adding up consecutive numbers. For example, the eighth triangular number is equal to –

1 + 2 + 3 + 4 + 5 + 6 + 7 + 8

– which comes to 36.

As the name suggests, you can visualise triangular numbers as a triangle of points.

Everyone in a group of people shakes hands with everyone else. The total number of handshakes will always be a triangular number. For instance, five people will make ten handshakes.

There is a useful short cut if you want to work out a large triangular number. Suppose you want the 100th triangular number. You could add up all the numbers from 1 to 100. But there is a simpler way. First work out the average number by adding together the first and the last number, and dividing by two –

1 + 100 = 101
101 / 2 = 50.5

Now multiply this average by however many numbers you would have to add up, in this case, by 100 –

50.5 x 100 = 5050

– so the 100th triangular number is 5050.

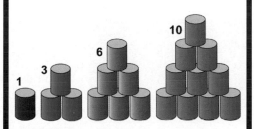

The first four triangular numbers. How many do you have to add each time?

140

= 2 x 2 x 5 x 7

= 2 x 70

= 4 x 35

= 5 x 28

= 7 x 20

= 10 x 14

= 1 + 4 + 9 + 16 + 25 + 36 + 49

A pyramidal number – the sum of the first seven square numbers.

The interior angles in a regular nine-sided figure (a nonagon) measure 140°.

On an oven, gas mark 1 means a temperature of 140° Celsius.

■ The great fictional detective Sherlock Holmes claimed to be able to identify 140 different types of tobacco from their ash.

In *The Sign of the Four* by Arthur Conan Doyle (1859 – 1930) we are told that Holmes had written an article 'Upon the Distinction between the Ashes of the Various Tobaccos' which listed 140 types of cigar, cigarette and pipe tobacco with coloured plates illustrating the differences in the ash.

PHOTO: JOHN GILLESPIE

Pyramids of oranges like this were once a common sight at greengrocers. There were several different ways of building them. For example, you could start with a 7 x 7 square of oranges, on top of this add a 6 x 6 square, followed by a 5 x 5 square, and so on, ending with a single orange on top of the pile. This way you would use exactly 140 oranges.

141

= 3 x 47

= 21 + 22 + 23 + 24 + 25 + 26

142

= 2 x 71

= 34 + 35 + 36 + 37

143

= 11 x 13

The new prisoner was puzzled because his fellow inmates laughed whenever one of them called out a number. He was told that the numbers were a code for jokes. To save time, you no longer needed to tell a joke, you just called out its number. The new prisoner was intrigued and decided to try it himself.

When the right moment came he called out '41' but was greeted with total silence. Back in his cell another prisoner explained where he had gone wrong. 'With jokes you have to remember it all depends on the way you tell them.'

(This is joke number 143.)

144
= 2 x 2 x 2 x 2 x 3 x 3
= 2 x 72
= 3 x 48
= 4 x 36
= 6 x 24
= 8 x 18
= 9 x 16
= 12 x 12
A square number.
= 1 + 3 + 5 + 7 + 9 + 11 + 13 + ... + 23

1, 1, 2, 3, 5, 8, 13...144
A Fibonacci number. It has been proved that 1 and 144 are the only square Fibonacci numbers. Curiously, 1 is the first Fibonacci number and 144 is the twelfth.

$12^2 = 144$ and reversing the digits – $21^2 = 441$.

The interior angles in a regular ten-sided figure (a decagon) measure 144°.

144 is called a *gross*. Many everyday objects like pins and buttons used to be sold by the gross. Half a gross of soap would mean 72 tablets of soap. If you wanted to count larger numbers than a gross you could use a *great gross* which equals twelve gross. This is part

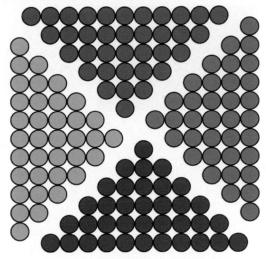

144 is a square number. Divide it by four and you get 36 which is also square. Does this happen with any other square numbers?

of a system of counting based on powers of twelve –
A dozen = 12
A gross = 12 x 12 = 144
A great gross = 12 x 12 x 12 = 1728

If this system is continued, what would the next unit be after a great gross? Work out its value and invent a name for it.

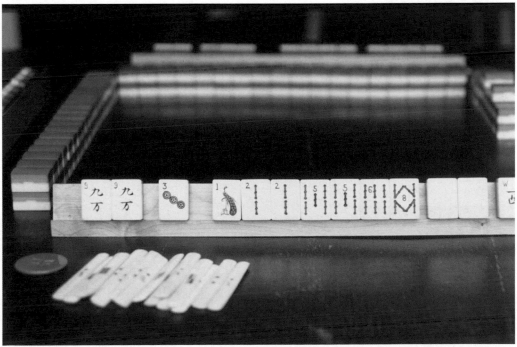

A full mah-jong set has 144 tiles.

. . . .

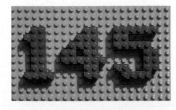

145
= 5 x 29
= 1! + 4! + 5!

There are only four numbers that equal the sum of the factorials of their digits. The other three are 1, 2 and 40,585.

146
= 2 x 73
= 35 + 36 + 37 + 38

146 is an octahedral number.
Each layer is a square:
1 + 4 + 9 + 16 + 25
+ 36
+ 25 + 16 + 9 + 4 + 1
= 146.

147
= 3 x 7 x 7
= 3 x 49
= 7 x 21

■ 147 is the maximum break in snooker. It has been achieved by more than 200 players. The first was by E J O'Donoghue in 1934 in New South Wales.

147 is the score for sinking 15 red balls scoring 1 point each, a yellow ball scoring 2 points, a green scoring 3, a brown scoring 4, a blue scoring 5, a pink scoring 6, and for scoring 7 points on each of the 16 occasions that the black ball is potted.

In fact it is possible to gain a higher score than 147 in snooker if your opponent makes a foul shot and you gain a free ball. In 1988 Steve Duggan achieved a score of 148 in this way. The maximum possible score, which has never been achieved, is 155.

148
= 2 x 2 x 37
= 2 x 74
= 4 x 37
= 15 + 16 + 17 + 18 + 19 + ... + 22

149 is a prime number.

149 rabbits are busy eating their way through a field of cabbages. The farmer raises his shot gun and fires, killing one of the rabbits. How many rabbits remain in the field?

150
= 2 x 3 x 5 x 5
= 2 x 75
= 3 x 50
= 5 x 30
= 6 x 25
= 10 x 15
= 7 + 8 + 9 + 10 + 11 + 12 + ... + 18

150 per cent of something means the same as 'one and half times'.

The brightest light bulbs you can normally buy are 150 watts.

On an oven, gas mark 2 is a temperature of 150° Celsius.

The internal angles in a regular 12 sided figure (a dodecagon) measure 150°.

A *sesquicentenary* is a fancy name for a hundred and fiftieth anniversary. 'Sesqui-' means one and a half, so a sesquicentenary celebrates 1½ x 100 = 150 years.

According to legend King Arthur had a round table, around which his 150 knights sat. How big would Arthur's table have been? Can you estimate its diameter?

151 is a prime number.

The gestation period of the goat is 151 days.

You can score 151 at darts by throwing two bulls and a treble 17. Is it possible to make 151 in a different way?

152
= 2 x 2 x 2 x 19
= 2 x 76
= 4 x 38
= 8 x 19
= 12 + 14 + 16 + 18 + 20 + ... + 26

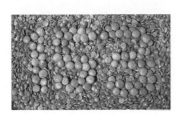

153
= 3 x 3 x 17
= 3 x 51
= 9 x 17
= 1 + 2 + 3 + 4 + 5 + 6 + ... + 17
A triangular number.

= $1^3 + 5^3 + 3^3$
The smallest number (after 1) which equals the sum of the cubes of its digits.

= 1! + 2! + 3! + 4! + 5!
The sum of the first five factorials.

■ In the Bible, the net which Simon Peter drew out of the sea at Tiberias contained 153 fishes (John 21:11). Why 153? Why not a nice round number like 100 or 150?

Numerologists have jumped at the opportunity for inventing ingenious symbolic explanations for the number 153. Saint Augustine discussed it at length but perhaps the definitive interpretation appears in the Reverend Bramley-Moore's *The Significance of Numbers as used in the Bible* (1917) which squeezes spiritual significance out of 153 and several dozen more numbers mentioned in the Bible.

As a result 153 has acquired a mystical status, and while it is not quite in the same league as the notorious 666, or the cosmic 42, it is perhaps not a number to be taken lightly.

154
= 2 x 7 x 11
= 2 x 77
= 7 x 22
= 11 x 14
= 9 + 10 + 11 + 12 + 13 + ... + 19

If you take a regular figure with nine sides (a nonagon) and join together all its points, you divide it into 154 regions. What happens if you do the same with other regular figures? Joining the corners of a square would produce just four regions. What happens with a regular pentagon? A regular hexagon?

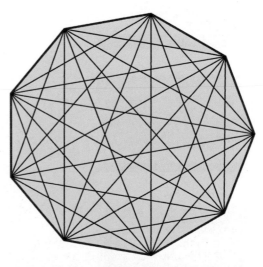

154 regions made by joining the points of a regular nonagon.

155

= 5 x 31

= 11 + 12 + 13 + 14 + 15 + ... + 20

If you built a house of cards like this but 10 storeys high it would contain 155 cards. You would need three packs of 52 playing cards and a lot of skill and patience. When you finished you would have one spare card.

In 1992 a new world record was set for building a house of cards using standard playing cards. Bryan Berg built a house of cards 75 storeys high. If he used the same method shown in this picture, how many cards would he have used?

Hint - try some simpler cases first.

156

= 2 x 2 x 3 x 13

= 2 x 78

= 3 x 52

= 4 x 39

= 6 x 26

= 12 x 13

A clock that strikes the hour will strike 156 times in the course of a day. It strikes once at one o'clock in the morning, twice at two o'clock, and so on up to 12 strikes at midday, making 78 strikes in all. The same number of strikes will occur from one o'clock in the afternoon until midnight, making a total of 156.

George Orwell's book *1984* begins 'It was a bright cold day in April, and the clocks were striking thirteen.'

It is of course inconceivable that a government would pass a law that clocks must strike the time according to the 24-hour clock. No one could possibly be that stupid, could they? But just in case, how many times would a 24-hour clock strike during a day?

157 is a prime number.

158

= 2 x 79

= 38 + 39 + 40 + 41

= 35 + 38 + 41 + 44

159

= 3 x 53

= 52 + 53 + 54

= 19 + 22 + 25 + 28 + 31 + 34

160
= 2 x 2 x 2 x 2 x 2 x 5
= 2 x 80
= 4 x 40
= 5 x 32
= 8 x 20
= 10 x 16
= 30 + 31 + 32 + 33 + 34

There are 160 fluid ounces in a gallon because 20 fluid ounces make a pint and eight pints make a gallon.

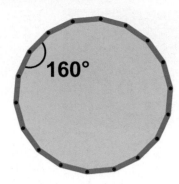

A regular figure with 18 sides has interior angles measuring 160°.

161
= 7 x 23
= 20 + 21 + 22 + 23 + 24 + 25 + 26
= 17 + 19 + 21 + 23 + 25 + 27 + 29

162
= 2 x 3 x 3 x 3 x 3
= 2 x 81
= 3 x 54
= 6 x 27
= 9 x 18
= 53 + 54 + 55

A regular figure with 20 sides (an icosagon) has interior angles measuring 162°.

An arrangement of 162 stars.

163 is a prime number.

The smallest score impossible to make with three darts. Other impossible scores are 166, 169, 172, 173, 175, 176, 178 and 179.

A pattern that uses 163 circles.

164
= 2 x 2 x 41
= 2 x 82
= 4 x 41
= 17 + 18 + 19 + 20 + 21 + ... + 24
= 10 + 13 + 16 + 19 + 22 + ... + 31

165

= 3 x 5 x 11

= 3 x 55

= 5 x 33

= 11 x 15

= 1 + 3 + 6 + 10 + 15 + ... + 45

The sum of the first nine triangular numbers and therefore a tetrahedral number.

A regular figure with 24 sides has interior angles measuring 165°.

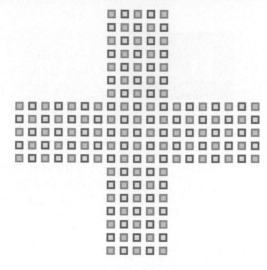

A cross made from 165 squares.

166

= 2 x 83

= 40 + 41 + 42 + 43

= 37 + 40 + 43 + 46

167 is a prime number. If you reverse its digits you get 761 which is also a prime number. Numbers like this are called *emirps* – emirp is the word prime backwards. The sequence of emirps begins 13, 17, 31, 37, 71, 73, 79, 97, 107, 113, 149, 157, 167...

Can you see why there are no emirps between 37 and 71?

168

= 2 x 2 x 2 x 3 x 7

= 2 x 84

= 3 x 56

= 4 x 42

= 6 x 28

= 7 x 24

= 8 x 21

= 12 x 14

If you think there are never enough hours in the week, you will not want to know that a week has 168 hours.

The total number of spots on a normal set of 28 dominoes is 168.

A regular figure with 30 sides has interior angles measuring 168°.

■ There are 168 prime numbers between one and one thousand. As numbers get larger, primes become less common –

4 primes from 1 to 10   (40%)

25 primes from 1 to 100   (25%)

168 primes from 1 to 1,000   (17%)

1,229 primes from 1 to 10,000   (12%)

9,592 primes from 1 to 100,000   (10%)

78,498 primes from 1 to 1,000,000   (8%)

Although the frequency of primes diminishes as numbers get bigger, it has been proved that there are an infinite number of prime numbers so the percentage never falls to zero.

169
= 13 x 13
A square number that connects to other square numbers.

$13^2 = 169$ and reversing both numbers it is also true that $31^2 = 961$.

169 is the sum of two squares:
$$169 = 5^2 + 12^2,$$
the sum of three squares:
$$169 = 3^2 + 4^2 + 12^2,$$
the sum of four squares:
$$169 = 4^2 + 5^2 + 8^2 + 8^2,$$
the sum of five squares:
$$169 = 3^2 + 4^2 + 4^2 + 8^2 + 8^2,$$
the sum of six squares:
$$169 = 2^2 + 2^2 + 5^2 + 6^2 + 6^2 + 8^2,$$
the sum of seven squares:
$$169 = 2^2 + 2^2 + 3^2 + 4^2 + 6^2 + 6^2 + 8^2$$
... and it doesn't stop there!

170
= 2 x 5 x 17
= 2 x 85
= 5 x 34
= 17 x 10
= 41 + 42 + 43 + 44

171
= 3 x 3 x 19
= 3 x 57
= 9 x 19
= 1 + 2 + 3 + 4 + 5 + 6 + 7 + ... + 18
A triangular number.
= 56 + 57 + 58
A regular figure with 40 sides has interior angles measuring 171°.

172
= 2 x 2 x 43
= 2 x 86
= 4 x 43

A pizza can be cut into 172 pieces with just 18 straight cuts.

A regular figure with 45 sides has interior angles measuring 172°.

173 is a prime number.

174
= 2 x 3 x 29
= 2 x 87
= 3 x 58
= 6 x 29
= 57 + 58 + 59
= 24 + 26 + 28 + 30 + 32 + 34

A regular figure with 60 sides has interior angles measuring 174°.

If you say that a Cadbury's Creme Egg has 174 'calories' you actually mean its energy content is 174 kcal. To the scientist this means that if you set fire to the chocolate egg, it would generate enough heat to raise the temperature of 174 kg of water by 1° Celsius. To everyone else, it means that you are going to put on weight if you eat a lot of them. Here are the calorific values of all sorts of things which you would not find in a health shop —

One Bassett's Jelly Baby	20 kcal
One McVities Jaffa Cake	48 kcal
One Jacobs Club Orange Biscuit	110 kcal
One Penguin Biscuit	131 kcal
One 34.5g packet Walkers Cheese & Onion crisps	181 kcal
One 62.5g Mars Bar	281 kcal
One 175g Terry's Chocolate Orange	927 kcal

175
= 5 x 5 x 7
= 5 x 35
= 7 x 25
= 33 + 34 + 35 + 36 + 37

175 is the constant of a 7 x 7 magic square.

$175 = 4^4 - 3^4$, the difference between two fourth powers.

A regular figure with 72 sides has interior angles measuring 175°.

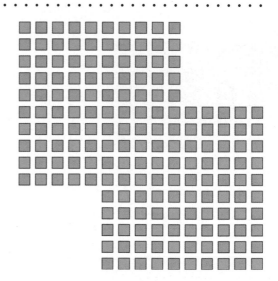

A pattern that uses 175 squares.

176
= 2 x 2 x 2 x 2 x 11
= 2 x 88
= 4 x 44
= 8 x 22
= 11 x 16
= 11 + 12 + 13 + 14 + 15 + ... + 21

176 is the 8th star number (or octagonal number). It is the 11th pentagonal number.

With ten straight cuts through a potato it is possible to divide it into 176 tiny pieces.

A regular figure with 90 sides has interior angles measuring 176°.

177
= 3 x 59
= 58 + 59 + 60
= 22 + 25 + 28 + 31 + 34 + 37

A regular figure with 120 sides has interior angles measuring 177°.

178
= 2 x 89
= 43 + 44 + 45 + 46
= 40 + 43 + 46 + 49

179 is a prime number.

Sophie Germain primes are calculated by taking a prime number, doubling it and adding one. For example, starting with 89 you get –
89, 179, 359, 719, 1439, 2879 ...
All of these numbers are primes but unfortunately the next number in the sequence is 5759 which equals 13 x 443 and so cannot be prime. In fact, any formula of this kind breaks down sooner or later. No one has found a formula that will generate a unending sequence of prime numbers. These special sequences of prime numbers are named after the French mathematician Marie-Sophie Germain (1776 – 1831).

180
= 2 x 2 x 3 x 3 x 5
= 2 x 90
= 3 x 60
= 4 x 45
= 5 x 36
= 6 x 30
= 9 x 20
= 10 x 18
= 12 x 15

180 is the smallest number to have 18 different factors, They are 1, 2, 3, 4, 5, 6, 9, 10, 12, 15, 18, 20, 30, 36, 45, 60, 90 and 180.

180 is the highest possible score with three darts. You need to hit three treble twenties.

180° is the sum of the interior angles of any triangle.

If you turn and face the opposite way you are turning through an angle of 180°.

On the Fahrenheit temperature scale ice melts at 32° and water boils at 212°, a difference of 180°.

A cassette marked 180 plays for three hours.

181 is a prime number.

181 is the tenth number that stays the same when written upside down.

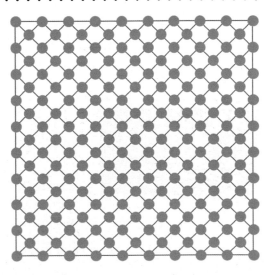

A pattern with 181 circles. One way to interpret this is to see that there is a 9 x 9 square within a 10 x 10 square.

182
= 2 x 7 x 13
= 2 x 91
= 7 x 26
= 44 + 45 + 46 + 47

Blink 182 is a US band formed in 1992. Their music has been described as 'melodic punk'.

183
= 3 x 61
= 28 + 29 + 30 + 31 + 32 + 33

184
= 2 x 2 x 2 x 23
= 2 x 92
= 4 x 46
= 8 x 23
= 31 + 41 + 51 + 61

185
= 5 x 37
= 35 + 36 + 37 + 38 + 39

186
= 2 x 3 x 31
= 2 x 93
= 3 x 62
= 6 x 31
= 26 + 28 + 30 +32 + 34 + 36

187
= 11 x 17
= 12 + 13 + 14 + 15 + 16 + ... + 22

The Secret Lives of Numbers website searches the internet to discover the popularity of different numbers by counting their occurrences on webpages. Sadly, in their 2002 survey, 187 was the least popular number under 200. The least popular under 100 was number 78. Not surprisingly, 1 is the most popular number, followed by other single digit numbers. In general, numbers ending 5 or 0 are more popular than those with other final digits, although there are many exceptions such as 12, 98 and 99.

According to the Bible (Genesis 5:25), when Methuselah was one hundred and eighty-seven years old he became the father of Lamech. He went on to live to an age of 969. These improbable achievements have given us the expression 'as old as Methuselah' meaning very old indeed.

188
= 2 x 2 x 47
= 2 x 94
= 4 x 47
= 44 + 46 + 48 + 50

This pattern is in a London pavement. It has 188 glass
blocks that allow daylight to reach the basement below.

189
= 3 x 3 x 3 x 7
= 3 x 63
= 7 x 27
= 9 x 21
= 29 + 30 + 31 + 32 + 33 + 34

189 is called a heptagonal number. It is
one of the sequence -
    1, 7, 18, 34, 55, 81, 112,
    148, 189, 235, 286, 342 ...
By taking differences it is easier to see
how the sequence is made –
    6, 11, 16, 21, 26, 31, 36 ...

190
= 2 x 5 x 19
= 2 x 95
= 5 x 38
= 10 x 19
= 1 + 2 + 3 + 4 + 5 + 6 ... + 19
  A triangular number.

190 is a hexagonal number – not to be
confused with the hex numbers. It is
part of the sequence –
    1, 6, 15, 28, 45, 66, 91, 120,
    153, 190, 231, 276, 325 ...
Taking differences you get 5, 9, 13,
17, 21, 25, 29 ... which go up in steps
of four. All hexagonal numbers are also
triangular numbers.

191 is a prime number.
There were 191 member states of the
United Nations (UN) in 2003. Each
member state has one vote. It might
seem odd that a tiny country like
Liechtenstein has the same voting
rights as a large country like India. But
in practice, all important UN decisions
are made by the UN Security Council
which has just 5 permanent members
(US, UK, France, Russian Federation
and China) and 10 elected members
drawn from the other countries. Any
one of the five permanent members
can veto (or block) a proposal before
the Security Council.

192
= 2 x 2 x 2 x 2 x 2 x 2 x 3
= 2 x 96
= 3 x 64
= 4 x 48
= 6 x 32
= 8 x 24
= 12 x 16
= 27 + 29 + 31 + 33 + 35 + 37

193 is a prime number.

194
= 2 x 97
= 47 + 48 + 49 + 50

195
= 3 x 5 x 13
= 3 x 65
= 5 x 39
= 13 x 15
= 30 + 31 + 32 + 33 + 34 + 35

196
= 2 x 2 x 7 x 7
= 2 x 98
= 4 x 49
= 7 x 28
= 14 x 14
A square number. Curiously 169, 196 and 961 are all square numbers.

■ A palindromic number is a number like 376,673 which reads the same backwards and forwards. Most numbers can be turned into palindromic numbers using a simple rule –

1 Choose any number,
2 Reverse its digits and add this to the original number,
3 If this is not a palindromic number go back to step 2 and repeat.

This palindrome-generating rule works for nearly every number. For example, applying it to 13 produces the palindrome 44, and 29 produces 121, both in one step. It takes two steps to convert 76 to 484. 80% of numbers under 10,000 produce a palindrome in 4 steps or fewer. The first number which appears to break the rule is 196. Starting with 196 and applying the rule again and again, you get larger and larger numbers but never get a palindrome. This has been tested on a computer by applying the rule more than 2 million times.

197 is a prime number.

198
= 2 x 3 x 3 x 11
= 2 x 99
= 3 x 66
= 6 x 33
= 9 x 22
= 11 x 18
= 28 + 30 + 32 + 34 + 36 + 38

199 is a prime number.

1, 3, 4, 7, 11, 18, 29 ... 199
A Lucas number.

199 countries were represented at the Sydney Olympic Games held in September and October 2000. This is the most countries who have participated in the games. The greatest number at a Winter Olympics was 77 countries at Utah in 2002.

200
= 2 x 2 x 2 x 5 x 5
= 2 x 100
= 4 x 50
= 5 x 40
= 8 x 25
= 10 x 20
= 20 + 30 + 40 + 50 + 60

The 200 euro note used in 12 countries.

Many races cover a distance of 200 metres. Swimming events over 200 metres include freestyle, butterfly, backstroke, breaststroke and medley.

■ In January 2001 the people in 12 European countries began using the common European currency. New coins and notes were issued including the 200 euro note shown here. Each country issues its own euro coins but the notes have the same design throughout Europe. The 200 euro note was designed by the Austrian artist Robert Kalina who had the problem of finding a design that would be acceptable in all 12 countries.

To remember the names of the 12 countries use the acronym BAFFLING PIGS: Belgium, Austria, Finland, France, Luxembourg, Ireland, the Netherlands, Germany, Portugal, Italy, Greece and Spain. If the UK, Denmark and Sweden also join the euro, a new acronym will be needed.

House numbers

Methods for numbering houses vary from place to place. In some roads number 13 is missing because someone thought it would be unlucky.

Very large house numbers, like 5926 in the picture, usually come about because some of the lower numbers have been missed out. In this road there was often a difference of eight between the number of one house and the next.

What do you do when extra houses are built between number 3 and number 4? The commonest solution is to call them 3A, 3B, 3C etc. In France they usually go for 3 *bis*, 3 *ter* etc. But in this street *(right)* they have used the fractions 3¼, 3½ and 3¾. This method used to be quite common in the UK but is rare today.

These two doors were in Quebec in Canada.

This pyramid is built entirely from pieces of Turkish delight. The base of the pyramid is made from a square 16 by 16, using 256 pieces of Turkish delight. On top of this is a second square 15 by 15, a third square 14 by 14, and so on, up to the single piece of Turkish delight at the top. The total number of pieces altogether is the sum of the first 16 square numbers –

$$1 + 4 + 9 + 16 + 25 \ldots + 256$$
$$= 1496 \text{ pieces.}$$

From 201 to 999

201 = 3 x 67
202 = 2 x 101
203 = 7 x 29
204 = 2 x 2 x 3 x 17, a pyramidal number.
205 = 5 x 41
206 = 2 x 103; adults normally have 206 bones in their bodies.
207 = 3 x 3 x 23
208 = 2 x 2 x 2 x 2 x 13
209 = 11 x 19
210 = 2 x 3 x 5 x 7, a triangular number, the product of the first four prime numbers.
211 is a prime number.
212 = 2 x 2 x 53; 212°F is water's boiling point.
213 = 3 x 71
214 = 2 x 107
215 = 5 x 43
216 = 2 x 2 x 2 x 3 x 3 x 3, a cube.
217 = 7 x 31
218 = 2 x 109
219 = 3 x 73
220 = 2 x 2 x 5 x 11, a tetrahedral number, an amicable number with 284; 220 yards = 1 furlong; when you join the points of a decagon there are 220 regions.
221 = 13 x 17
222 = 2 x 3 x 37
223 is a prime number.
224 = 2 x 2 x 2 x 2 x 2 x 7
225 = 3 x 3 x 5 x 5, a square, a star number.
226 = 2 x 113
227 is a prime number.
228 = 2 x 2 x 3 x 19
229 is a prime number.
230 = 2 x 5 x 23
231 = 3 x 7 x 11, a triangular number.
232 = 2 x 2 x 2 x 29
233 is a prime number, a Fibonacci number.
234 = 2 x 3 x 3 x 13
235 = 5 x 47
236 = 2 x 2 x 59
237 = 3 x 79
238 = 2 x 7 x 17
239 is a prime number.
240 = 2 x 2 x 2 x 2 x 3 x 5; about 240 human beings are born every minute; 240 old pennies = £1; 240 chains = 1 league.
241 is a prime number.
242 = 2 x 11 x 11
243 = 3 x 3 x 3 x 3 x 3, a fifth power, the Cockcroft Number.
244 = 2 x 2 x 61
245 = 5 x 7 x 7
246 = 2 x 3 x 41
247 = 13 x 19
248 = 2 x 2 x 2 x 31
249 = 3 x 83
250 = 2 x 5 x 5 x 5
251 is a prime number.
252 = 2 x 2 x 3 x 3 x 7
253 = 11 x 23, a triangular number.
254 = 2 x 127
255 = 3 x 5 x 17
256 = 2 x 2 x 2 x 2 x 2 x 2 x 2 x 2, a square, a fourth power, an eighth power.
257 is a prime number.
258 = 2 x 3 x 43

259 = 7 x 37
260 = 2 x 2 x 5 x 13, the constant of an 8 x 8 magic square.
261 = 3 x 3 x 29
262 = 2 x 131
263 is a prime number.
264 = 2 x 2 x 2 x 3 x 11
265 = 5 x 53
266 = 2 x 7 x 19
267 = 3 x 89; 267 days is the average human gestation period
268 = 2 x 2 x 67
269 is a prime number.
270 = 2 x 3 x 3 x 3 x 5
271 is a prime number.
272 = 2 x 2 x 2 x 2 x 17
273 = 3 x 7 x 13, 273°K is the melting point of ice.
274 = 2 x 137
275 = 5 x 5 x 11
276 = 2 x 2 x 3 x 23, a triangular number.
277 is a prime number.
278 = 2 x 139
279 = 3 x 3 x 31
280 = 2 x 2 x 2 x 5 x 7, a star number.
281 is a prime number.
282 = 2 x 3 x 47
283 is a prime number.
284 = 2 x 2 x 71, an amicable number with 220.
285 = 3 x 5 x 19, a pyramidal number.
286 = 2 x 11 x 13, a tetrahedral number.
287 = 7 x 41
288 = 2 x 2 x 2 x 2 x 2 x 3 x 3
289 = 17 x 17, a square.
290 = 2 x 5 x 29
291 = 3 x 97
292 = 2 x 2 x 73
293 is a prime number.
294 = 2 x 3 x 7 x 7
295 = 5 x 59
296 = 2 x 2 x 2 x 37
297 = 3 x 3 x 3 x 11
298 = 2 x 149
299 = 13 x 23

300

300 = 2 x 2 x 3 x 5 x 5, a triangular number.
301 = 7 x 43
302 = 2 x 151
303 = 3 x 101
304 = 2 x 2 x 2 x 2 x 19
305 = 5 x 61
306 = 2 x 3 x 3 x 17
307 is a prime number.
308 = 2 x 2 x 7 x 11
309 = 3 x 103

310 = 2 x 5 x 31
311 is a prime number.
312 = 2 x 2 x 2 x 3 x 13
313 is a prime number.
314 = 2 x 157
315 = 3 x 3 x 5 x 7
316 = 2 x 2 x 79
317 is a prime number.
318 = 2 x 3 x 53
319 = 11 x 29
320 = 2 x 2 x 2 x 2 x 2 x 2 x 5; 320 rods = 1 mile.
321 = 3 x 107
322 = 2 x 7 x 23, a Lucas number.
323 = 17 x 19
324 = 2 x 2 x 3 x 3 x 3 x 3, a square.
325 = 5 x 5 x 13, a triangular number.
326 = 2 x 163
327 = 3 x 109
328 = 2 x 2 x 2 x 41
329 = 7 x 47
330 = 2 x 3 x 5 x 11
331 is a prime number.
332 = 2 x 2 x 83
333 = 3 x 3 x 37; there are 333 different heptahexes.
334 = 2 x 167
335 = 5 x 67
336 = 2 x 2 x 2 x 2 x 3 x 7
337 is a prime number.
338 = 2 x 13 x 13
339 = 3 x 113
340 = 2 x 2 x 5 x 17
341 = 11 x 31, a star number.
342 = 2 x 3 x 3 x 19
343 = 7 x 7 x 7, a cube.
344 = 2 x 2 x 2 x 43
345 = 3 x 5 x 23
346 = 2 x 173
347 is a prime number.
348 = 2 x 2 x 3 x 29
349 is a prime number.
350 = 2 x 5 x 5 x 7
351 = 3 x 3 x 3 x 13, a triangular number.
352 = 2 x 2 x 2 x 2 x 2 x 11

353 is a prime number; international dialling code for Republic of Ireland.
354 = 2 x 3 x 59
355 = 5 x 71
356 = 2 x 2 x 89
357 = 3 x 7 x 17
358 = 2 x 179
359 is a prime number.
360 = 2 x 2 x 2 x 3 x 3 x 5; in a revolution there are 360°.
361 = 19 x 19, a square; a go board has 361 positions.
362 = 2 x 181
363 = 3 x 11 x 11
364 = 2 x 2 x 7 x 13, a tetrahedral number.
365 = 5 x 73; there are 365 days in a normal year.
366 = 2 x 3 x 61; there are 366 days in a leap year.
367 is a prime number.
368 = 2 x 2 x 2 x 2 x 23
369 = 3 x 3 x 41, the constant of a 9 x 9 magic square.
370 = 2 x 5 x 37
371 = 7 x 53
372 = 2 x 2 x 3 x 31
373 is a prime number.
374 = 2 x 11 x 17
375 = 3 x 5 x 5 x 5
376 = 2 x 2 x 2 x 47
377 = 13 x 29, a Fibonacci number.
378 = 2 x 3 x 3 x 3 x 7, a triangular number.
379 is a prime number.

Amicable numbers

The two smallest amicable numbers are 220 and 284.

The factors of 220, apart from itself, are 1, 2, 4, 5, 10, 11, 20, 22, 44, 55 and 110. These add up to 284.

The factors of 284, apart from itself are 1, 2, 4, 71 and 142. And these add up to 220.

A pair of numbers are amicable if the factors of one number (excluding itself) add up to the other number. The adding up is done in the same way as for perfect numbers.

After 220 and 284, the next pair of amicable numbers are 1184 and 1210.

380 = 2 x 2 x 5 x 19
381 = 3 x 127
382 = 2 x 191
383 is a prime number.
384 = 2 x 2 x 2 x 2 x 2 x 2 x 2 x 3
385 = 5 x 7 x 11, a pyramidal number.
386 = 2 x 193
387 = 3 x 3 x 43
388 = 2 x 2 x 97
389 is a prime number.
390 = 2 x 3 x 5 x 13
391 = 17 x 23
392 = 2 x 2 x 2 x 7 x 7
393 = 3 x 131
394 = 2 x 197
395 = 5 x 79
396 = 2 x 2 x 3 x 3 x 11
397 is a prime number.
398 = 2 x 199
399 = 3 x 7 x 19

400

400 = 2 x 2 x 2 x 2 x 5 x 5, a square.
401 is a prime number.
402 = 2 x 3 x 67
403 = 13 x 31
404 = 2 x 2 x 101
405 = 3 x 3 x 3 x 3 x 5; in UK, old TV pictures had 405 lines.
406 = 2 x 7 x 29, a triangular number.
407 = 11 x 37
408 = 2 x 2 x 2 x 3 x 17, a star number.
409 is a prime number.
410 = 2 x 5 x 41
411 = 3 x 137
412 = 2 x 2 x 103
413 = 7 x 59
414 = 2 x 3 x 3 x 23
415 = 5 x 83
416 = 2 x 2 x 2 x 2 x 2 x 13
417 = 3 x 139
418 = 2 x 11 x 19
419 is a prime number.
420 = 2 x 2 x 3 x 5 x 7
421 is a prime number.
422 = 2 x 211
423 = 3 x 3 x 47
424 = 2 x 2 x 2 x 53
425 = 5 x 5 x 17
426 = 2 x 3 x 71
427 = 7 x 61
428 = 2 x 2 x 107
429 = 3 x 11 x 13
430 = 2 x 5 x 43
431 is a prime number.
432 = 2 x 2 x 2 x 2 x 3 x 3 x 3
433 is a prime number.
434 = 2 x 7 x 31
435 = 3 x 5 x 29, a triangular number.
436 = 2 x 2 x 109
437 = 19 x 23
438 = 2 x 3 x 73
439 is a prime number.
440 = 2 x 2 x 2 x 5 x 11; in music the A above middle C has a pitch of 440 Hz.
441 = 3 x 3 x 7 x 7, a square.
442 = 2 x 13 x 17
443 is a prime number.
444 = 2 x 2 x 3 x 37
445 = 5 x 89
446 = 2 x 223
447 = 3 x 149
448 = 2 x 2 x 2 x 2 x 2 x 2 x 7
449 is a prime number.
450 = 2 x 3 x 3 x 5 x 5
451 = 11 x 41
452 = 2 x 2 x 113
453 = 3 x 151
454 = 2 x 227
455 = 5 x 7 x 13, a tetrahedral number.
456 = 2 x 2 x 2 x 3 x 19
457 is a prime number.
458 = 2 x 229
459 = 3 x 3 x 3 x 17
460 = 2 x 2 x 5 x 23
461 is a prime number.
462 = 2 x 3 x 7 x 11

463 is a prime number.
464 = 2 x 2 x 2 x 2 x 29
465 = 3 x 5 x 31, a triangular number.
466 = 2 x 233
467 is a prime number.
468 = 2 x 2 x 3 x 3 x 13
469 = 7 x 67
470 = 2 x 5 x 47
471 = 3 x 157
472 = 2 x 2 x 2 x 59
473 = 11 x 43
474 = 2 x 3 x 79
475 = 5 x 5 x 19
476 = 2 x 2 x 7 x 17
477 = 3 x 3 x 53
478 = 2 x 239
479 is a prime number.
480 = 2 x 2 x 2 x 2 x 2 x 3 x 5
481 = 13 x 37, a star number.
482 = 2 x 241
483 = 3 x 7 x 23
484 = 2 x 2 x 11 x 11, a square.
485 = 5 x 97
486 = 2 x 3 x 3 x 3 x 3 x 3
487 is a prime number.
488 = 2 x 2 x 2 x 61
489 = 3 x 163
490 = 2 x 5 x 7 x 7
491 is a prime number.
492 = 2 x 2 x 3 x 41
493 = 17 x 29
494 = 2 x 13 x 19
495 = 3 x 3 x 5 x 11
496 = 2 x 2 x 2 x 2 x 31, a triangular number, the third perfect number.
497 = 7 x 71
498 = 2 x 3 x 83
499 is a prime number.

500

500 = 2 x 2 x 5 x 5 x 5; the Romans wrote 500 as D, 'a monkey'.
501 = 3 x 167; darts is often played by subtracting scores from 501.
502 = 2 x 251
503 is a prime number.
504 = 2 x 2 x 2 x 3 x 3 x 7
505 = 5 x 101, the constant of a 10 x 10 magic square.
506 = 2 x 11 x 23, a pyramidal number.
507 = 3 x 13 x 13
508 = 2 x 2 x 127
509 is a prime number.
510 = 2 x 3 x 5 x 17
511 = 7 x 73
512 = 2 x 2 x 2 x 2 x 2 x 2 x 2 x 2 x 2, a cube, a ninth power.
513 = 3 x 3 x 3 x 19
514 = 2 x 257
515 = 5 x 103
516 = 2 x 2 x 3 x 43
517 = 11 x 47
518 = 2 x 7 x 37
519 = 3 x 173
520 = 2 x 2 x 2 x 5 x 13
521 is a prime number, a Lucas number.
522 = 2 x 3 x 3 x 29
523 is a prime number.
524 = 2 x 2 x 131
525 = 3 x 5 x 5 x 7
526 = 2 x 263
527 = 17 x 31
528 = 2 x 2 x 2 x 2 x 3 x 11, a triangular number.
529 = 23 x 23, a square.
530 = 2 x 5 x 53
531 = 3 x 3 x 59
532 = 2 x 2 x 7 x 19
533 = 13 x 41
534 = 2 x 3 x 89
535 = 5 x 107
536 = 2 x 2 x 2 x 67
537 = 3 x 179
538 = 2 x 269
539 = 7 x 7 x 11
540 = 2 x 2 x 3 x 3 x 3 x 5; the interior angles of a pentagon add up to 540°.

541 is a prime number.
542 = 2 x 271
543 = 3 x 181
544 = 2 x 2 x 2 x 2 x 2 x 17
545 = 5 x 109
546 = 2 x 3 x 7 x 13
547 is a prime number.
548 = 2 x 2 x 137
549 = 3 x 3 x 61
550 = 2 x 5 x 5 x 11
551 = 19 x 29
552 = 2 x 2 x 2 x 3 x 23
553 = 7 x 79
554 = 2 x 277
555 = 3 x 5 x 37
556 = 2 x 2 x 139
557 is a prime number.
558 = 2 x 3 x 3 x 31
559 = 13 x 43
560 = 2 x 2 x 2 x 2 x 5 x 7, a tetrahedral number, a star number.
561 = 3 x 11 x 17, a triangular number.
562 = 2 x 281
563 is a prime number.
564 = 2 x 2 x 3 x 47
565 = 5 x 113
566 = 2 x 283
567 = 3 x 3 x 3 x 3 x 7
568 = 2 x 2 x 2 x 71
569 is a prime number.
570 = 2 x 3 x 5 x 19
571 is a prime number.
572 = 2 x 2 x 11 x 13
573 = 3 x 191
574 = 2 x 7 x 41
575 = 5 x 5 x 23
576 = 2 x 2 x 2 x 2 x 2 x 2 x 3 x 3, a square.
577 is a prime number.
578 = 2 x 17 x 17
579 = 3 x 193
580 = 2 x 2 x 5 x 29
581 = 7 x 83
582 = 2 x 3 x 97
583 = 11 x 53
584 = 2 x 2 x 2 x 73
585 = 3 x 3 x 5 x 13
586 = 2 x 293
587 is a prime number.
588 = 2 x 2 x 3 x 7 x 7
589 = 19 x 31
590 = 2 x 5 x 59
591 = 3 x 197
592 = 2 x 2 x 2 x 2 x 37
593 is a prime number.
594 = 2 x 3 x 3 x 3 x 11
595 = 5 x 7 x 17, a triangular number.
596 = 2 x 2 x 149
597 = 3 x 199
598 = 2 x 13 x 23
599 is a prime number.

600

600 = 2 x 2 x 2 x 3 x 5 x 5
601 is a prime number.
602 = 2 x 7 x 43
603 = 3 x 3 x 67
604 = 2 x 2 x 151
605 = 5 x 11 x 11
606 = 2 x 3 x 101
607 is a prime number.
608 = 2 x 2 x 2 x 2 x 2 x 19
609 = 3 x 7 x 29; stays the same when written upside down.
610 = 2 x 5 x 61, a Fibonacci number.
611 = 13 x 47
612 = 2 x 2 x 3 x 3 x 17
613 is a prime number.
614 = 2 x 307
615 = 3 x 5 x 41
616 = 2 x 2 x 2 x 7 x 11
617 is a prime number.
618 = 2 x 3 x 103
619 is a prime number; stays the same when written upside down.
620 = 2 x 2 x 5 x 31
621 = 3 x 3 x 3 x 23
622 = 2 x 311
623 = 7 x 89

624 = 2 x 2 x 2 x 2 x 3 x 13
625 = 5 x 5 x 5 x 5, a square, a fourth power, European TV pictures have 625 lines.
626 = 2 x 313
627 = 3 x 11 x 19
628 = 2 x 2 x 157
629 = 17 x 37
630 = 2 x 3 x 3 x 5 x 7, a triangular number.
631 is a prime number.
632 = 2 x 2 x 2 x 79
633 = 3 x 211
634 = 2 x 317
635 = 5 x 127
636 = 2 x 2 x 3 x 53
637 = 7 x 7 x 13
638 = 2 x 11 x 29
639 = 3 x 3 x 71
640 = 2 x 2 x 2 x 2 x 2 x 2 x 2 x 5; 640 acres = 1 square mile.
641 is a prime number.
642 = 2 x 3 x 107
643 is a prime number.
644 = 2 x 2 x 7 x 23
645 = 3 x 5 x 43, a star number.
646 = 2 x 17 x 19
647 is a prime number.
648 = 2 x 2 x 2 x 3 x 3 x 3 x 3
649 = 11 x 59
650 = 2 x 5 x 5 x 13, a pyramidal number.
651 = 3 x 7 x 31
652 = 2 x 2 x 163
653 is a prime number.
654 = 2 x 3 x 109
655 = 5 x 131
656 = 2 x 2 x 2 x 2 x 41
657 = 3 x 3 x 73
658 = 2 x 7 x 47
659 is a prime number.
660 = 2 x 2 x 3 x 5 x 11
661 is a prime number.
662 = 2 x 331
663 = 3 x 13 x 17
664 = 2 x 2 x 2 x 83
665 = 5 x 7 x 19
666 = 2 x 3 x 3 x 37, a triangular number, the 'Number of the Beast' in the Bible.
667 = 23 x 29
668 = 2 x 2 x 167
669 = 3 x 223
670 = 2 x 5 x 67
671 = 11 x 61, the constant of an 11 x 11 magic square.
672 = 2 x 2 x 2 x 2 x 2 x 3 x 7
673 is a prime number.
674 = 2 x 337
675 = 3 x 3 x 3 x 5 x 5
676 = 2 x 2 x 13 x 13, a square.
677 is a prime number.
678 = 2 x 3 x 113
679 = 7 x 97
680 = 2 x 2 x 2 x 5 x 17, a tetrahedral number.
681 = 3 x 227
682 = 2 x 11 x 31
683 is a prime number.
684 = 2 x 2 x 3 x 3 x 19
685 = 5 x 137
686 = 2 x 7 x 7 x 7
687 = 3 x 229
688 = 2 x 2 x 2 x 2 x 43
689 = 13 x 53; stays the same when written upside down.
690 = 2 x 3 x 5 x 23
691 is a prime number.
692 = 2 x 2 x 173
693 = 3 x 3 x 7 x 11
694 = 2 x 347
695 = 5 x 139
696 = 2 x 2 x 2 x 3 x 29
697 = 17 x 41
698 = 2 x 349
699 = 3 x 233

700

700 = 2 x 2 x 5 x 5 x 7
701 is a prime number.

702 = 2 x 3 x 3 x 3 x 13
703 = 19 x 37, a triangular number.
704 = 2 x 2 x 2 x 2 x 2 x 2 x 11
705 = 3 x 5 x 47
706 = 2 x 353
707 = 7 x 101
708 = 2 x 2 x 3 x 59
709 is a prime number.
710 = 2 x 5 x 71
711 = 3 x 3 x 79
712 = 2 x 2 x 2 x 89
713 = 23 x 31
714 = 2 x 3 x 7 x 17
715 = 5 x 11 x 13
716 = 2 x 2 x 179
717 = 3 x 239
718 = 2 x 359
719 is a prime number.
720 = 2 x 2 x 2 x 2 x 3 x 3 x 5; a factorial (6!); the interior angles of a hexagon add up to 720°; has 30 factors.
721 = 7 x 103
722 = 2 x 19 x 19
723 = 3 x 241
724 = 2 x 2 x 181
725 = 5 x 5 x 29
726 = 2 x 3 x 11 x 11
727 is a prime number.
728 = 2 x 2 x 2 x 7 x 13
729 = 3 x 3 x 3 x 3 x 3 x 3, a square, a cube, a sixth power.
730 = 2 x 5 x 73
731 = 17 x 43
732 = 2 x 2 x 3 x 61
733 is a prime number.
734 = 2 x 367
735 = 3 x 5 x 7 x 7
736 = 2 x 2 x 2 x 2 x 2 x 23, a star number.
737 = 11 x 67
738 = 2 x 3 x 3 x 41
739 is a prime number.
740 = 2 x 2 x 5 x 37
741 = 3 x 13 x 19, a triangular number.
742 = 2 x 7 x 53
743 is a prime number.
744 = 2 x 2 x 2 x 3 x 31
745 = 5 x 149
746 = 2 x 373
747 = 3 x 3 x 83; the Boeing 747 aeroplane is the well known 'jumbo jet'.
748 = 2 x 2 x 11 x 17
749 = 7 x 107
750 = 2 x 3 x 5 x 5 x 5
751 is a prime number.
752 = 2 x 2 x 2 x 2 x 47
753 = 3 x 251
754 = 2 x 13 x 29
755 = 5 x 151
756 = 2 x 2 x 3 x 3 x 3 x 7
757 is a prime number.
758 = 2 x 379
759 = 3 x 11 x 23
760 = 2 x 2 x 2 x 5 x 19
761 is a prime number.
762 = 2 x 3 x 127
763 = 7 x 109
764 = 2 x 2 x 191
765 = 3 x 3 x 5 x 17
766 = 2 x 383
767 = 13 x 59
768 = 2 x 2 x 2 x 2 x 2 x 2 x 2 x 2 x 3
769 is a prime number.
770 = 2 x 5 x 7 x 11
771 = 3 x 257
772 = 2 x 2 x 193
773 is a prime number.
774 = 2 x 3 x 3 x 43
775 = 5 x 5 x 31
776 = 2 x 2 x 2 x 97
777 = 3 x 7 x 37
778 = 2 x 389
779 = 19 x 41
780 = 2 x 2 x 3 x 5 x 13, a triangular number.
781 = 11 x 71
782 = 2 x 17 x 23

783 = 3 x 3 x 3 x 29
784 = 2 x 2 x 2 x 2 x 7 x 7, a square.
785 = 5 x 157
786 = 2 x 3 x 131
787 is a prime number.
788 = 2 x 2 x 197
789 = 3 x 263
790 = 2 x 5 x 79
791 = 7 x 113
792 = 2 x 2 x 2 x 3 x 3 x 11
793 = 13 x 61
794 = 2 x 397
795 = 3 x 5 x 53
796 = 2 x 2 x 199
797 is a prime number.
798 = 2 x 3 x 7 x 19
799 = 17 x 47

800

800 = 2 x 2 x 2 x 2 x 2 x 5 x 5
801 = 3 x 3 x 89
802 = 2 x 401
803 = 11 x 73
804 = 2 x 2 x 3 x 67
805 = 5 x 7 x 23
806 = 2 x 13 x 31
807 = 3 x 269
808 = 2 x 2 x 2 x 101; stays the same when written upside down.
809 is a prime number.
810 = 2 x 3 x 3 x 3 x 3 x 5
811 is a prime number.
812 = 2 x 2 x 7 x 29
813 = 3 x 271
814 = 2 x 11 x 37
815 = 5 x 163
816 = 2 x 2 x 2 x 2 x 3 x 17, a tetrahedral number.
817 = 19 x 43
818 = 2 x 409; stays the same when written upside down.
819 = 3 x 3 x 7 x 13, a pyramidal number.
820 = 2 x 2 x 5 x 41, a triangular number.
821 is a prime number.
822 = 2 x 3 x 137
823 is a prime number.
824 = 2 x 2 x 2 x 103
825 = 3 x 5 x 5 x 11
826 = 2 x 7 x 59
827 is a prime number.
828 = 2 x 2 x 3 x 3 x 23
829 is a prime number.
830 = 2 x 5 x 83
831 = 3 x 277
832 = 2 x 2 x 2 x 2 x 2 x 2 x 13
833 = 7 x 7 x 17, a star number.
834 = 2 x 3 x 139
835 = 5 x 167
836 = 2 x 2 x 11 x 19
837 = 3 x 3 x 3 x 31
838 = 2 x 419
839 is a prime number.
840 = 2 x 2 x 2 x 3 x 5 x 7; has 32 factors – the most in a number less than 1,000.
841 = 29 x 29, a square.
842 = 2 x 421
843 = 3 x 281, a Lucas number.
844 = 2 x 2 x 211
845 = 5 x 13 x 13
846 = 2 x 3 x 3 x 47
847 = 7 x 11 x 11
848 = 2 x 2 x 2 x 2 x 53
849 = 3 x 283
850 = 2 x 5 x 5 x 17
851 = 23 x 37
852 = 2 x 2 x 3 x 71
853 is a prime number.
854 = 2 x 7 x 61
855 = 3 x 3 x 5 x 19
856 = 2 x 2 x 2 x 107
857 is a prime number.
858 = 2 x 3 x 11 x 13
859 is a prime number.
860 = 2 x 2 x 5 x 43
861 = 3 x 7 x 41, a triangular number.
862 = 2 x 431
863 is a prime number.

864 = 2 x 2 x 2 x 2 x 2 x 3 x 3 x 3
865 = 5 x 173
866 = 2 x 433
867 = 3 x 17 x 17
868 = 2 x 2 x 7 x 31
869 = 11 x 79
870 = 2 x 3 x 5 x 29, the constant of a 12 x 12 magic square.
871 = 13 x 67
872 = 2 x 2 x 2 x 109
873 = 3 x 3 x 97, = 1! + 2! + 3! + 4! + 5! + 6!
874 = 2 x 19 x 23
875 = 5 x 5 x 5 x 7
876 = 2 x 2 x 3 x 73
877 is a prime number.
878 = 2 x 439
879 = 3 x 293
880 = 2 x 2 x 2 x 2 x 5 x 11
881 is a prime number.
882 = 2 x 3 x 3 x 7 x 7
883 is a prime number.
884 = 2 x 2 x 13 x 17
885 = 3 x 5 x 59
886 = 2 x 443
887 is a prime number.
888 = 2 x 2 x 2 x 3 x 37; stays the same when written upside down; in UK teletext number for television subtitles for the hard of hearing.
889 = 7 x 127
890 = 2 x 5 x 89
891 = 3 x 3 x 3 x 3 x 11
892 = 2 x 2 x 223
893 = 19 x 47
894 = 2 x 3 x 149
895 = 5 x 179
896 = 2 x 2 x 2 x 2 x 2 x 2 x 2 x 7
897 = 3 x 13 x 23
898 = 2 x 449
899 = 29 x 31

900

900 = 2 x 2 x 3 x 3 x 5 x 5, a square; the interior angles of a heptagon add up to 900°.
901 = 17 x 53
902 = 2 x 11 x 41
903 = 3 x 7 x 43, a triangular number.
904 = 2 x 2 x 2 x 113
905 = 5 x 181
906 = 2 x 3 x 151; stays the same when written upside down.
907 is a prime number.
908 = 2 x 2 x 227
909 = 3 x 3 x 101
910 = 2 x 5 x 7 x 13
911 is a prime number; phone number for emergency services in USA; the date 11 September remembered in Chile for 1973 and in the USA for 2001.
912 = 2 x 2 x 2 x 2 x 3 x 19
913 = 11 x 83
914 = 2 x 457
915 = 3 x 5 x 61
916 = 2 x 2 x 229; stays the same when written upside down.
917 = 7 x 131
918 = 2 x 3 x 3 x 3 x 17
919 is a prime number.
920 = 2 x 2 x 2 x 5 x 23
921 = 3 x 307
922 = 2 x 461
923 = 13 x 71
924 = 2 x 2 x 3 x 7 x 11
925 = 5 x 5 x 37
926 = 2 x 463
927 = 3 x 3 x 103
928 = 2 x 2 x 2 x 2 x 2 x 29
929 is a prime number.
930 = 2 x 3 x 5 x 31
931 = 7 x 7 x 19
932 = 2 x 2 x 233
933 = 3 x 311
934 = 2 x 467
935 = 5 x 11 x 17
936 = 2 x 2 x 2 x 3 x 3 x 13, a star number.
937 is a prime number.

938 = 2 x 7 x 67
939 = 3 x 313
940 = 2 x 2 x 5 x 47
941 is a prime number.
942 = 2 x 3 x 157
943 = 23 x 41
944 = 2 x 2 x 2 x 2 x 59
945 = 3 x 3 x 3 x 5 x 7
946 = 2 x 11 x 43, a triangular number.
947 is a prime number.
948 = 2 x 2 x 3 x 79
949 = 13 x 73
950 = 2 x 5 x 5 x 19
951 = 3 x 317
952 = 2 x 2 x 2 x 7 x 17
953 is a prime number.
954 = 2 x 3 x 3 x 53
955 = 5 x 191
956 = 2 x 2 x 239
957 = 3 x 11 x 29
958 = 2 x 479
959 = 7 x 137
960 = 2 x 2 x 2 x 2 x 2 x 2 x 3 x 5; 960 rods = 1 league; has 28 factors.
961 = 31 x 31, a square.
962 = 2 x 13 x 37
963 = 3 x 3 x 107
964 = 2 x 2 x 241
965 = 5 x 193
966 = 2 x 3 x 7 x 23
967 is a prime number.
968 = 2 x 2 x 2 x 11 x 11
969 = 3 x 17 x 19, a tetrahedral number.
970 = 2 x 5 x 97
971 is a prime number.
972 = 2 x 2 x 3 x 3 x 3 x 3 x 3
973 = 7 x 139
974 = 2 x 487
975 = 3 x 5 x 5 x 13
976 = 2 x 2 x 2 x 2 x 61
977 is a prime number.
978 = 2 x 3 x 163
979 = 11 x 89
980 = 2 x 2 x 5 x 7 x 7
981 = 3 x 3 x 109
982 = 2 x 491
983 is a prime number.
984 = 2 x 2 x 2 x 3 x 41
985 = 5 x 197
986 = 2 x 17 x 29; stays the same when written upside down.
987 = 3 x 7 x 47, a Fibonacci number.
988 = 2 x 2 x 13 x 19
989 = 23 x 43
990 = 2 x 3 x 3 x 5 x 11, a triangular number.
991 is a prime number.
992 = 2 x 2 x 2 x 2 x 2 x 31
993 = 3 x 331
994 = 2 x 7 x 71
995 = 5 x 199
996 = 2 x 2 x 3 x 83
997 is a prime number.
998 = 2 x 499
999 = 3 x 3 x 3 x 37, phone number for emergency services in the UK.

A few large numbers

666 666 is 'The Number of the Beast', the devil's number, which comes directly from chapter 13 of the Book of Revelation in the Bible –

> And I stood upon the sand of the sea and saw a beast rise up out of the sea, having seven heads and ten horns, and upon his horns ten crowns, and upon his heads the name of blasphemy.
>
> ...and power was given unto him to continue forty and two months...
>
> Here is wisdom. Let him that hath understanding count the number of the beast: for it is the number of a man; and his number is six hundred threescore and six.

This deliciously sinister piece of mysticism has caught the imagination of many people during the 2,000 years since it was written and there have been many interpretations of what it is about.

Some theories latch on to 'the number of a man' and assume that a particular person is involved. The Roman emperor Nero is one possibility. Perhaps the writer was afraid to name him directly and had to refer to him in code? If you translate Nero's name into Hebrew you get NRON KSR. Next you substitute numbers for letters according to a particular system of numerology and add these up –

$$50 + 200 + 6 + 50 + 100 + 60 + 200 = 666.$$

If you are not convinced by this argument you can choose between dozens of other historical figures such as Martin Luther, Muhammad and Napoleon Bonaparte, all of whom have been candidates for the beast.

A few years ago, crazy advertisements appeared in Australian newspapers warning people about the bar codes used on ordinary consumer goods –

> This bar code system on all merchant goods is for easy identification and control using a laser computer incorporating the international 666 system. Be warned that the next world dictator as predicted by the Bible is here and this is how we will know him. As all goods are numbered so shortly will all people be numbered with an invisible laser tattoo on the right hand or the forehead...

Although less sensational, there is a much simpler explanation for the choice of 666 in the Bible. Expressed in Roman numerals, 666 becomes DCLXVI. This number uses each of the Roman numerals in turn and has an appealing pattern to it like 123456. It is possible that for the writer of the Book of Revelation 666 was just a convenient large number.

The Romans wrote 1,000 as M, which is an abbreviation for their word *mille*, meaning a thousand.

A millennium is 1,000 years and a millipede has 1,000 legs – or at least it did in the imagination of the person who named it. A distance of one mile was originally 1,000 paces – the Romans counted one pace as two steps.

With units of measurement, *milli-* means one thousandth. So 1,000 millimetres make one metre and 1,000 milligrams make one gram. *Milli-* is abbreviated to *m*, e.g. mW means milliwatt.

Kilo- means a thousand times so 1,000 metres make a kilometre. *Kilo-* is shortened to *k*, e.g. kg means kilogram.

'The face that launched a thousand ships' according to the playwright Christopher Marlowe (1564 – 1593) belonged to Helen of Troy. Her legendary beauty is supposed to have caused the war between the Greeks and the Trojans. Recently this has led to the suggestion that beauty should be measured in a unit called the *millihelen*. A millihelen is defined as the amount of beauty that would launch a single ship.

The number 1,729 has been made famous by a meeting between two mathematicians – Ramanujan and G H Hardy. The brilliant Indian mathematician Ramanujan had an uncanny ability to remember details about numbers. Hardy recalls on one occasion –

> I remember going to see him once when he was lying ill in Putney. I had ridden in a taxi-cab No.1,729, and remarked that the number seemed to me a rather dull one, and that I hoped it was not an unfavourable omen. 'No,' he reflected, 'it is a very interesting number; it is the smallest number expressible as the sum of two cubes in two different ways.'

From his remarkable knowledge, Ramanujan had instantly recalled that –
$$10^3 + 9^3 = 1{,}729$$
and also that –
$$12^3 + 1^3 = 1{,}729$$

· ·

42,195 metres (or 26 miles 385 yards) is now the recognised distance of a marathon race. This gruelling race was introduced to the 1896 Olympic Games to commemorate a famous event in ancient Greece: Philippides' run from the battlefield of Marathon to Athens in 490 BC. As no one knows the exact distance he ran, marathons have been run over varying distances, but 42,195 metres is now recognised as the Olympic standard. Before 1970, the marathon was a men's event, but now both women and men compete in marathons.

· ·

One hundred thousand is also called a *lakh*, which is a Hindu word.

Ex -ASP held over Rs. 5 lakh heist

by **Bandula Dinapurna, Norman Palihawadana** and **Sarath Chandrasiri**
A former Assistant Superintendent of Police was arrested yesterday (25) by Mirihana police in connection with a rupee five lakh robbery which took place at Nugegoda last Thursday (24).

The word lakh is frequently used on the Indian subcontinent. This cutting is from a newspaper in Sri Lanka and is about the theft of 500,000 (5 lakh) rupees.

Imagining large numbers

The number displayed on a computer screen begins at zero. Every time someone jumps on the black mat, it goes up by one.

If it takes half a second to jump once, you can move the counter to '10' in five seconds. You can move it to '100' in 50 seconds. But what about larger numbers?

Imagine a team of people taking turns to jump on the mat. Working day and night, they keep up a rate of one jump every half second.

1,000 jumps take
8 minutes 20 seconds,

10,000 jumps take
1 hour 23 minutes,

100,000 jumps take
13 hours 53 minutes,

1,000,000 jumps take
5 days 19 hours,

10,000,000 jumps take
8 weeks 2 days,

100,000,000 jumps take
1 year 31 weeks,

1,000,000,000 jumps take
15 years 44 weeks.

1000000

A million is a square number, a cube and a sixth power.

A millionairess or millionaire is someone with a wealth of a million pounds, a million dollars, or a million of some other currency.

The Million Pound Note is a story by Mark Twain about a man who is given a million pound banknote which he finds difficult to spend.

Mega- means a million times so 1,000,000 metres make a megametre. *Mega-* is abbreviated to 'M', e.g. MHz means megahertz.

Micro- means one millionth so 1,000,000 micrometres make one metre. The micrometre is also called a *micron. Micro-* is abbreviated to the Greek letter 'µ', which is pronounced 'mew', e.g. µg means microgram.

Names for large numbers

Number	Zeros	In USA called –	In Europe called –	Metric prefix	Metric prefix for fraction
10	1	Ten	Ten	Hecto- *	Deci- *
100	2	Hundred	Hundred	Deca- or Deka- *	Centi- *
1,000	3	Thousand	Thousand	Kilo- e.g. kilogram (kg)	Milli- e.g. millisecond (ms)
1,000,000	6	Million	Million	Mega- e.g. megawatt (MW)	Micro- e.g. microvolt (µV)
1,000,000,000	9	Billion	Milliard	Giga- e.g. gigabyte (Gbyte) **	Nano- e.g. nanolitre (nl)
1,000,000,000,000	12	Trillion	Billion	Tera- e.g. terametre (Tm)	Pico- e.g. picofarad (pF)
1,000,000,000,000,000	15	Quadrillion	Thousand billion ***	Peta- e.g. petahertz (PHz)	Femto- e.g. femtojoule (fJ)
1,000,000,000,000,000,000	18	Quintillion	Trillion	Exa- e.g. exapascal (EPa)	Atto- e.g. attonewton (aN)
1,000,000,000,000,000,000,000	21	Sextillion	Thousand trillion		
1,000,000,000,000,000,000,000,000	24	Septillion	Quadrillion		
1,000,000,000,000,000,000,000,000,000	27	Octillion	Thousand quadrillion		
1,000,000,000,000,000,000,000,000,000,000	30	Nonillion	Quintillion		
1,000,000,000,000,000,000,000,000,000,000,000	33	Decillion	Thousand quintillion		

* Centi-, deci-, deca-, and hecto- are widely used in units like the centimetre and decilitre, but scientists and engineers tend to discourage the use of these four prefixes.

** A kilobyte, which measures the amount of memory in a computer, is only roughly equal to 1000 bytes –

1 kilobyte = 1024 bytes (2^{10})
1 megabyte = 1048576 bytes (2^{20})
1 gigabyte = 1073741824 bytes (2^{30})

*** Logically this should be called a 'billiard'?

1,000,000,000 is called a *billion* in the USA but it is called a *milliard* (or a thousand million) in most parts of Europe. The European billion (10^{12}) is a thousand times larger than the American billion (10^9).

In the United Kingdom chaos rules because both meanings of a billion are used, although the American usage is now more common. Government departments have been known to switch from one usage to the other and letters appear from time to time in the British press defending the European billion from what is seen as an unnecessary Americanisation.

Like a billion, the names trillion, quadrillion etc. also have different meanings and so need to be used with care. However a *zillion* remains a useful word for an indefinitely large number wherever you happen to be.

1.496×10^{11}

149,600,000,000 metres is the distance which astronomers call one astronomical unit (au). It is the average distance between the Earth and the sun.

Astronomers also measure huge distances in light years. This is the distance that light travels through space in one year. 63240 astronomical units roughly equal one light year which is about 9,460,500,000,000,000 metres.

The Googol is a number invented by the nine-year-old nephew of mathematician Edward Kasner. It is simply –

10,000,000,000,000,000,000,000,000,
000,000,000,000,000,000,000,000,000,
000,000,000,000,000,000,000,000,000,
000,000,000,000,000,000,000

– or 1 followed by a hundred zeros (10^{100}).

And if you think that is a big number, then a googolplex is mind-bogglingly bigger. It was invented at the same time as a googol. A googolplex is 1 followed by a googol of zeros. It is so large that it could never be written down in digits.

Both numbers are in the realm of mathematical fantasy. Neither could ever be used to count anything in the real world. It has been estimated that there are only about 10^{80} particles in the entire universe, which comes a long way short of a googol. But large numbers like these still have their uses in mathematics. The largest known prime number is much larger than a googol but insignificantly small in comparison to a googolplex.

Mathematicians use the symbol ∞ as an abbreviation for infinity.

This book is mostly about the counting numbers 1, 2, 3, 4 ... What is the highest counting number? This question really does not make sense because we know they go on for ever – give me a number as big as you like and I can always give you a bigger one. But it makes more sense to ask how many counting numbers there are altogether, and this is one definition of infinity.

Infinity is not really a number and to treat it like a number makes nonsense of ordinary arithmetic. If you add 12 to ∞ you still have ∞.

If you stand between two facing mirrors, you will see images of yourself repeated many, many times. You could start to count these images but you know they go on for ever, even though they get fainter and fainter as they get further away. According to this definition, infinity is something that is countable but goes on for ever.

How many even numbers are there? Because half of the counting numbers are even numbers you might think the answer is ∞/2, but the answer is in fact still ∞. This is because you can match up the even numbers with the counting numbers –

    2 goes with 1,
    4 goes with 2,
    6 goes with 3,
    8 goes with 4,
    10 goes with 5,
– and this must go on for ever.

This definition of infinity means that anything you can match up in this way, one-to-one, with the counting numbers is itself infinite in quantity. So if there are an infinite number of counting numbers, there are also an infinite number of prime numbers, of square numbers, of numbers that contain a '6', of palindromic numbers, and so on.

If this is not confusing enough, mathematicians have invented a second definition of infinity, which must be larger than the first definition, because when you try to do this matching up, you find that there are objects left over. One example is the set of real numbers. These are numbers that can have any number of digits after the decimal point (e.g. 3.333... or π 3.14159...). It is not possible to match these up with the counting numbers and so they are an example of this second definition of infinity.

This limerick was written by the physicist George Gamow who was struck by the paradox of a finite being contemplating infinity –

    There was a young fellow from Trinity
    Who took the square root of infinity.
    But the number of digits
    Gave him the fidgets;
    He dropped maths and took up divinity.

Between two facing mirrors the images go on for ever.

The idea of infinity is more familiar than most people realise.

The top picture shows the design painted on the side of an arcade game machine. You can see a circle with a square drawn inside it. Inside that square is another circle, and inside that circle is another square. The pattern could go on for ever using smaller and smaller squares and circles. How many squares or circles would be used? The answer is of course infinity.

The middle picture shows a camera lens. It has a scale for focusing showing the distance in feet ... 10, 15, 30, ∞. In photography, any distance much greater than 30 feet can be thought of as infinitely far away.

The bottom picture shows a railway line. We know that the rails never meet. But in the photograph they appear to meet at the top of the picture, in the far distance. It is sometimes helpful to think of parallel lines meeting at an infinite distance.

Hints, answers & more questions

Not all the questions posed in this book are answered. When a problem is in blue lettering, you may find a matching entry here that provides a hint, or even an answer. In some cases there is an additional question to work on.

0

When you enter 6 ÷ 0 most calculators display 'E' meaning there is an error. They are unable to do the calculation for you.

Try dividing 6 by numbers that are closer and closer to zero, such as –
0.1, 0.01, 0.001 ...
– and you will find the answer gets larger and larger –
60, 600, 6000...

As you get closer to zero, the answer gets impossibly large, and mathematicians say 'it tends towards infinity'. But infinity (∞) is not really a number and so the rules of arithmetic simply say that you are not allowed to divide by zero.

If you break this rule you risk getting answers that make no sense. For example, it is true that –
5 x 0 = 6 x 0
– but if you throw caution to the winds and divide both sides of this equation by zero, you get –
5 = 6
– which is clearly not true.

2

There is no whole number with this property but if you experiment with a calculator, you will find that 1.732... gives the same answer when you add it or multiply it three times. 1.732... is the square root of 3.

3

Not three, but four chocolate bars fit into the box in the photograph. A triangular box that was one size larger would take nine bars. Can you see a pattern here and use it to predict the number of bars in still larger boxes?

4

As well as making a square and a parallelogram, you should be able to make a trapezium (called a trapezoid in the USA) and a kite shape. It is also possible to make any number of irregular quadrilaterals, including some concave quadrilaterals.

5

There are 12 different pentominoes. If you think you have found more than 12, you will find that two are the same when you rotate them or turn them over.

You can make an intriguing puzzle by cutting out the shapes of the 12 pentominoes and making one extra piece which is a 2 x 2 square (a tetromino). The problem is to assemble these into an 8 x 8 square like a chessboard.

Hexominoes and heptominoes are shapes made by joining six and seven squares respectively. There are 35 different hexominoes and 108 different heptominoes.

7 (Hexagons)

If you are having problems, try drawing out the hexagons on triangular paper. The paper makes it easy to draw patterns of regular hexagons.

How many shapes can be made by linking five regular hexagons?

7 (St Ives)

It is common to see someone analyse the problem in the following way.

There was 1 man,
there were 7 wives,
there were 7 x 7 = 49 sacks,
there were 7 x 7 x 7 = 343 cats,
there were 7 x 7 x 7 x 7 = 2401 kits.

Adding these up you get a total of 2801 going to St Ives.

But of course, this is not the right answer. Look at the rhyme again and you will see that the correct answer is much simpler to work out.

8

Here is one solution to the problem. The key is to think about the two numbers in the middle of the diagram. Both of these are joined to six other numbers. Only 1 and 8 can take up these middle positions.

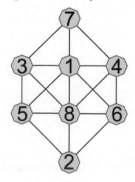

10 (Paris & Rome)

If you are stuck on this riddle, figure out what 509 is in Roman numbers.

10 (Paper sizes)

From an A0 sheet of paper you cut either –

    2 A1 sheets,
    4 A2 sheets,
    8 A3 sheets,
    16 A4 sheets,
    32 A5 sheets,
    64 A6 sheets,
    128 A7 sheets,
    256 A8 sheets,
    512 A9 sheets, or
    1024 A10 sheets.

One sheet of 80gsm A0 paper must weigh 80 grams because it has an area of one square metre. This A0 sheet can be cut into 16 A4 sheets which must also weight 80 grams.

11

69, 88 and 96 all stay the same when written upside down. Which numbers come next?

If you write 1999 upside down you get 6661. What was the last year which remained unchanged when written upside down? When is the next?

12 (Queens)

Here is one way in which 8 queens can be arranged on a chessboard.

What is the greatest number of knights that can be placed on a chess board in such a way that no knight attacks another?

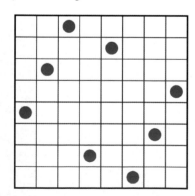

12 (Ruler)

If your ruler has as few marks as possible it will have a '0', a '6' and just two other marks.

13 (Friday)

A calendar should help you solve the first question.

If April 13 is a Friday then July 13 is also a Friday and it is the next one to occur.

In this case there is a gap of 91 days –

7 x 13 = 91

– this gap must always be a multiple of 7 because there are seven days in a week. But it must also equal the sum of the number of days in the months. April has 30 days, May has 31 and June 30 days, adding up to 91.

This may help you think about the other questions: what is the shortest possible interval and the longest possible interval?

13 (Farm shop)

A simpler case may help you see an answer to this. Suppose you had just two weights: 1 kg and 3 kg. With these you could weigh any whole number of kilograms from 1 kg to 4 kg. To weigh 2 kg you place the two weights on opposite sides of the scales. You then add potatoes to the lighter side until the scales balance. At this point there must be exactly 2 kg of potatoes on the scales.

When you have solved the problem with three weights, what fourth weight would allow you to weigh any whole number of kilograms up to 40 kg?

18

Using whole numbers, the only rectangles where the area equals the perimeter are 6 by 3 and 4 by 4. One way to prove this is to construct a spreadsheet to calculate the area minus the perimeter for rectangles of every size from 1 by 1 to 10 by 10.

20

If the wheel let you rotate just two numbers it would be much easier to put the numbers into order. One way would be to work round the ring considering each pair of numbers and rotating any pairs which are out of order. If you go round several times you will eventually get them into order.

But if the wheel lets you rotate three numbers the task becomes impossible. Imagine that the numbers around the ring are coloured alternately dark and light. If you rotated any three of them, the colours would still alternate dark and light. So it is impossible to sort them into a pattern of your choice.

21

The first six star numbers are 1, 8, 21, 40, 65 and 96. You can check your answer for the next one, which is larger than a hundred, by looking up its entry in this book. Star numbers are equal in value to what mathematicians call octagonal numbers. The star shape is a kind of octagon because it has eight sides.

24

'375' and the other modern numbers used to mark gold show the amount of gold in parts per thousand.

Pure gold is 24 carat and so 9 carat gold contains $^9/_{24}$ of pure gold. Expressed as a decimal this is 0.375. As a percentage it is 37.5 per cent, or 375 parts per thousand.

If you wanted to use a calculator to convert you would multiply the new numbers by 0.024, for example –

750 x 0.024 = 18 carat

But how could you use a calculator to convert from a carat number to the new numbers?

25

Jurassic Park would play for about 127 minutes in the cinema at 24 pictures a second. On television, if it was shown in full without breaks, it would play for about 122 minutes at 25 pictures a second, which is five minutes shorter.

If a velociraptor dinosaur runs at 20 mph in the cinema, will it appear to move slower or faster on television?

26

Here is the solution Dudeney gives. The sides of the heptagon, which each add up to 26 are –

14, 2, 10
10, 3, 13
13, 4, 9
9, 5, 12
12, 6, 8
8, 7, 11
11, 1, 14

Using the same diagram and the same numbers from 1 to 14, find an arrangement so that each side now adds up to 19.

It is also possible to find solutions where every side comes to 22 or to 23.

27

Almost everyone makes a mistake when they try this problem for the first time. It may help to make a sketch and mark off the cubes as you count them. If you are confident that you have a correct answer, check that your total number of cubes comes to 27 and that the total number of red faces is 54.

28

If you find that 28 of your digits do not make a cubit, do not assume that the ancient Egyptians got it wrong! Although body measurements were the basis for their measuring system, they used carefully standardised units as we do today. They needed to measure with great precision in order to build the pyramids.

30

A right-angled triangle with sides 6, 8 and 10 has an area and a perimeter both equal to 24.

32

Packs of 32 playing cards are used for card games like piquet and bezique. They consist of the Ace, King, Queen, Jack, 10, 9, 8 and 7 in all four suits.

36 (Market researcher)

The woman had twins aged 2 and an older child aged 9. The problem is delightful because it appears to not give you enough information. If you are stuck, try writing out all the different ages that multiply together to make 36. Then ask yourself why the salesman needed to ask for a final clue.

36 (Stamps)

The illustration shows how a 4 by 9 block of stamps can be torn into two pieces and rearranged as a 6 by 6 block. Here is another problem. How can you make a 3 by 12 block by tearing the original 4 by 9 block into just two pieces?

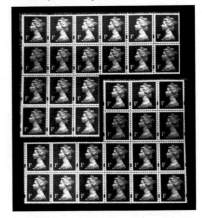

37

Just three colours are needed to ensure that no hexagon is next to another hexagon with the same colour. Here is one way to do it.

40 (Puck)

In 1988 an aircraft flew around the world in 36 hours 9 minutes with an average speed of 1026 km/h or 638 mph. Puck's fictional circumnavigation must have been about 60 times faster.

40 (Rods)

The old unit called a rod is about five metres in length which is the height of a tall giraffe.

41

How many tiles are used for different sizes?

A pattern 3 tiles across uses 5 tiles,
5 tiles across uses 13 tiles,
7 tiles across uses 25 tiles,
9 tiles across uses 41 tiles...

The numbers go 5, 13, 25, 41... How does this number pattern continue? Try taking differences between these numbers, or thinking about how many extra tiles are added on each time. A pattern 19 tiles across uses 181 tiles.

42

It is not difficult to figure out the number in the middle. The cube uses all the numbers from 1 to 27, and you know 26 of them, so which is missing?

44

How many paving slabs for a square pond?

A 2 x 2 pond needs 12 slabs,
a 3 x 3 pond needs 16 slabs,
a 4 x 4 pond needs 20 slabs,
a 5 x 5 pond needs 24 slabs ...

Can you see a rule emerging here?

What happens with ponds that are rectangular in shape? What happens with L-shaped ponds?

45

A mystic rose with 20 points around the circle would use 190 lines. Is it just a coincidence that both 45 and 190 are triangular numbers?

48 (Douzièmes)

A douzième measures a very tiny distance. It is approximately 0.18 mm which is about the thickness of two pages of this book.

48 (Triangles)

Surprisingly, there are more ways to make a triangle with 11 matches than with 12 matches.

With 11 matches there are four ways. You can make a 4-4-3 triangle, a 5-3-3 triangle, a 5-4-2 triangle or a 5-5-1 triangle.

But with 12 matches there are only three ways. You can make a 4-4-4 triangle, a 5-4-3 triangle or a 5-5-2 triangle.

What happens with other numbers of matches?

54

270 inches of cloth is the same as six English ells, or five French ells, or ten Flemish ells. It is the shortest length that gives an exact number in all three systems.

55

If you have a set of 55 dominoes available, try making a chain and see what happens.

If this is not an option for you, it may help to think about the dominoes with blanks on them. One tile is a double blank and there are nine other tiles which have a blank on one side. These must be put together in pairs but because nine is an odd number there will be one tile left over which is impossible to fit into the chain. With a normal set of 28 dominoes there are six tiles which have a blank on one side. Because this is an even number they can all be put together into pairs.

56

Think about a simpler case where only the numbers from 0 to 2 are used. The tiles would be 0-0-0, 0-0-1, 0-0-2, 0-1-1, 0-1-2, 0-2-2, 1-1-1, 1-1-2, 1-2-2 and 2-2-2 which makes ten in all.

Now what about tiles with numbers from 0 to 3? And is it just a coincidence that both 10 and 56 are tetrahedral numbers?

63

You may have spotted that there are in fact two answers to this question: 81 and 18 are one pair. 92 and 29 are the other.

65

There were 60 students at the meal. There are lots of ways of solving this kind of problem. The question tells you that the number of students must be exactly divisible by 2, 3 and 4. So it is a number in the twelve-times table. It is also clear that the number of students is roughly equal to the number of dishes, so 60 looks a good bet, and when you check this out, it fits.

66

Some people can solve this problem by imagining that the blocks are rearranged into a simpler pattern. A different approach is to consider some simpler cases and see how the number of blocks increases as the height goes up. Perhaps you have found a different way? For a tower 12 blocks high the answer is 276 blocks.

Try to find a formula to give the number of blocks for any size of skeleton tower.

71

Zoe could have claimed a total of ten free packets. To begin, she would have traded 64 empty packets for eight full ones. When she had eaten them all, the eight empty packets would be exchanged for another full one. When this was empty it would be combined with the seven packets (still remaining from the original 71) to claim one more free packet, making ten in all.

72

45 is the only number that is five times the sum of its own digits. Can you find a number which is seven times the sum of its own digits? There are four different answers. Try to find them all.

77

It is impossible to make a score of 77. Because both 9 and 6 are multiples of 3, any score must be a number in the three-times table. This rules out 77.

How close can you get to 77?

78

If you take the song literally –

1 present was given on the first day,
1 + 2 = 3 presents on the second day,
1 + 2 + 3 = 6 presents on the third day,
1 + 2 + 3 + 4 = 10 on the fourth day,
... and so on up to 78 presents on the twelfth day.

The grand total is –
1 + 3 + 6 + 10 + 15 + 21 + 28 + ... + 78
= 364 presents, or almost one for every day of the year.

You may recognise that the number on each of the days is a triangular number and that

the total of 364 must therefore be a tetrahedral number. Fancifully all the presents could be stacked in a gigantic tetrahedron with each triangular layer containing the presents for one of the days of Christmas.

There are other answers of course depending on how you interpret the song. We have assumed that a 'partridge in a pear tree' is one present, but if it counts as two, things would be different.

81

After the change 64 models would be available. There are eight possible front wheels and eight possible rear wheels making 8 x 8 = 64 combinations.

85

The next 3 lines of the pattern would be –

$15^2 + 112^2 = 113^2$ $15 + 112 - 113 = 14$
$17^2 + 144^2 = 145^2$ $17 + 144 - 145 = 16$
$19^2 + 180^2 = 181^2$ $19 + 180 - 181 = 18$

The odd numbers 3, 5, 7, 9... appear in the first column. If you take differences between the numbers in the second column you will find they go up in multiples of four: + 8, + 12, + 16, + 20 ... etc. The number in the third column is always one more than the number in the second column.

91

There are 30 squares altogether in a 4 x 4 pattern.

Here are some results for other sizes –
1 x 1 has 1 square,
2 x 2 has 5 squares,
3 x 3 has 14 squares,
4 x 4 has 30 squares,
5 x 5 has 55 squares,
6 x 6 has 91 squares,
7 x 7 has ...?

Can you see a pattern in these numbers? Use it to find the answer for other sizes of square.

93

A 3 x 3 square uses 9 circles, a 5 x 5 square uses 21 circles and a 7 x 7 square uses 33 circles. Because there must be a circle in the centre of the pattern, there must be an odd number of circles along each side.

96

With four skirts, eight tops and four pairs of shoes there are 4 x 8 x 4 = 128 different outfits.

How does this change if Jamilla decides that the colours of one skirt and one top clash, and so can never be worn together?

99

This is how the pattern grows –
$9^2 = 81$
$99^2 = 9,801$
$999^2 = 998,001$
$9,999^2 = 99,980,001$
$99,999^2 = 9,999,800,001$
$999,999^2 = 999,998,000,001$
$9,999,999^2 = 99,999,980,000,001$

105

The extra sock could be worn with any of the 15 original socks. So there are 15 more pairs, making 120 pairs altogether.

The problem assumes that it does not matter which way round you wear a pair of odd socks.

109

In Australia there are only four ways of paying 20c. These are 20c, 10c + 10c, 10c + 5c + 5c and 5c + 5c + 5c + 5c.

In the USA there are nine ways of paying 20c which are 10c + 10c, 10c + 5c + 5c, 10c + 5c + five 1c, 10c + ten 1c, 5c + 5c + 5c + 5c, 5c + 5c + 5c + five 1c, 5c + 5c + ten 1c, 5c + fifteen 1c and twenty 1c.

How about 25 cents? 30 cents?

111

Here is part of the pattern which of course goes on for ever –

$1^2 = 1$
$11^2 = 121$
$111^2 = 12,321$
$1,111^2 = 1,234,321$
$11,111^2 = 123,454,321$
$111,111^2 = 12,345,654,321$
$1,111,111^2 = 1,234,567,654,321$
$11,111,111^2 = 123,456,787,654,321$
$111,111,111^2 = 12,345,678,987,654,321$

It has been proved that no repunits, however long, can be square numbers.

Mathematicians have done a lot of research to discover which repunits are prime numbers. These first prime repunit is 11. The next is 19 digits long.

113

These are the two-digit prime numbers which are also prime when their digits are reversed –
13 and 31,
17 and 71,
37 and 73,
79 and 97,
and 11 which stays the same when reversed.

114

If you celebrate your 114th birthday during 2005 you must have been born in the year 1891, when Queen Victoria was still on the throne in England.

117

The diagram shows how three of the patterns fit together to make an equilateral triangle.

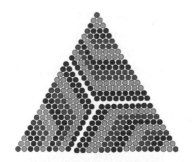

118

With four rows there are a total of 27 triangles. This is made up of 16 small triangles, 7 triangles of the next size (of which one is upside down), 3 still larger triangles, and one triangle formed by the whole pattern itself.

As the pattern grows larger the numbers of small triangles are square numbers of increasing size –
1, 4, 9, 16, 25, 36, 49 ...
– but there is not such a tidy pattern to the number of triangles of larger sizes. Here is how the total numbers increase –

Rows	Triangles
1	1
2	5
3	13
4	27
5	48
6	78
7	118
...	...

Here is a slightly easier question to think about. How many matchsticks would you need to make each pattern? One row would use three matches and two rows would use nine. How does this continue? Can you find a rule? And can you use this rule to figure out the number of matches needed to make the pattern with 100 rows?

120

10 also has the property of being both a triangular and a tetrahedral number. On small tables, snooker is sometimes played with 10 balls, rather than the usual 15.

126

She used the 2-litre measure three times, the 3-litre measure ten times, and the 5-litre measure 18 times.

131

If you have the answer 2,248,091 cubic metres, take another look at the problem.

139

Apart from 1, 7 is the lowest happy number. The sequence goes –
7, 49, 97, 130, 10, 1.

4 is an example of an unhappy number. It goes –
4, 16, 37, 58, 89, 145, 42, 20, 4 ...
Notice it has returned to 4 again. This sequence of eight numbers just goes on repeating for ever and so can never reach 1.

Can you find all the happy numbers less than 20? Besides 1 and 7 there are three others.

144

The next number after a great gross would be 12 x 12 x 12 x 12 = 20,736 which could perhaps be called a 'gross gross'.

149

The field would of course contain just one dead rabbit.

150

There is not enough information to give an exact answer to this question but it is not difficult to make an estimate.

We might guess that each knight occupied about 60 cm around the edge of the table. As there are 150 knights, the distance around the table (the circumference) would be –
150 x 60 = 9000 cm
– which is 90 metres.

You could work out the distance across the table (the diameter) by drawing a scale plan, but it is easier if you know the formula that tells you, for any circle –
Circumference = Diameter x π
– where π is the number called 'pi' equal to 3.14159...

So by dividing 90 by 3.141... we find that the diameter of Arthur's table must have been about 29 metres.

At Winchester there is a round table top that is said to be Arthur's one. Although it is very big, its diameter is not as large as 29 metres. Perhaps there were less knights or perhaps they just squeezed together?

151

Another way to make 151 with three darts would be to throw a double 20, a treble 20 and a treble 17. There is at least one other way to make 151. Can you find it?

154

If you use straight lines to join all the corners of a regular figure, this is the number of regions you make –

1 region with 3 sides (triangle),
4 regions with 4 sides (square),
11 regions with 5 sides (pentagon),
24 regions with 6 sides (hexagon),
50 regions with 7 sides (heptagon),
80 regions with 8 sides (octagon),
154 regions with 9 sides (nonagon),
220 regions with 10 sides (decagon).

Can you find a rule for predicting the number of regions? It is easier if you consider the odd and even numbers of sides separately.

155

There is more than one way to build a house of cards. Using the method shown in the photograph, a house of cards 75 storeys high would use 8475 cards. The easiest way to get this answer is to consider some simpler cases first and then look for a pattern –

2 cards with 1 storey,
7 cards with 2 storeys,
15 cards with 3 storeys ...

How does this continue? Can you discover a rule for any number of storeys?

In fact, the current world record holder used a different method of building from that shown in the photograph.

156

A 24 hour clock would chime –
1 + 2 + 3 + 4 + ... + 24
– or 300 times during the course of a day.

Further reading

Books about numbers

A Book of Numbers by John Grant. 1982, Ashgrove Press, 0 906798 19 1.

The Book of Numbers by John H Conway & Richard K Guy. 1996, Copernicus, 0 387979 93 X.

The Guinness Book of Numbers by Adrian Room. 1989, Guinness. 0 851123 72 4.

The Mystery of Numbers by Annemarie Schimmel. 1993, Oxford University Press, 0 195089 19 7.

The Number File by Adrian Jenkins. 2000, Tarquin Publications, 1 899618 40 6.

Number Words and Number Symbols: A Cultural History of Numbers by Karl Menninger, translated from the German by Paul Broneer. 1969, MIT Press, 0 262630 61 3.

Numbers: From Ancient Civilisations to the Computer by John McLeish. 1991, Flamingo, 0 006544 84 3.

Numbers: Their History and Meaning by Graham Flegg. 1983, Andre Deutsch, 0 233975 16 0.

The Penguin Dictionary of Curious and Interesting Numbers by David Wells. Revised edition 1997, Penguin, 0 140261 49 4.

The Universal History of Numbers by Georges Ifrah. 1998, Harvill, 1 860463 24 X.

Mathematics books

Game, Set and Math by Ian Stewart. 1989, Penguin, 0 140132 37 6.

How Puzzling by Charles Snape & Heather Scott. 1991, Cambridge University Press, 0 521356 73 3.

In Code: A Mathematical Journey by Sarah Flannery. 2001, Profile Books, 1 861972 71 7.

Mathematical Excursions by Helen A Merrill. 1933, reprinted as a Dover paperback, 0 486203 50 6.

Mathematical Magic Show by Martin Gardner. 1986, Penguin, 0 140165 56 8.

Mathematics: A Human Endeavour by Harold R Jacobs. 1982, Freeman, 0 716713 30 6.

Nature's Numbers by Ian Stewart. 1995, Weidenfeld & Nicolson. 0 297816 42 X.

The Number Detective by Jon Millington. 2001, Tarquin Publications, 1 899618 33 3.

A Number for your Thoughts by Malcolm E Lines. 1986, Institute of Physics Publishing, 0 852744 95 1.

Problems with Patterns and Numbers edited by Malcolm Swan. 1984, Shell Centre for Mathematical Education, 0 901628 32 8.

Think of a Number by Malcolm E Lines. 1990, Adam Hilger, 0 852741 83 9.

Why do Buses come in Threes? by Rob Eastaway & Jeremy Wyndham. 1998, Robson. 1 861051 62 X.

Websites

These website addresses were correct at the time of writing.

Maths Year 2000 Numberland. Number poems, limericks, facts, and pictures all behind a giant magic square.
www.mathsyear2000.co.uk/ numberland/

Numbers from one to thirty one. A page for each day of the month. Based on this book.
www.richardphillips.org.uk/number/

Numbers from 1 to 10 in over 4500 languages.
www.zompist.com/numbers.shtml

The Digits Project. Hundreds of facts about the digits from 0 to 9.
www-personal.umich.edu/%7Ebrinck/ digits/digits.html

The Secret Life of Numbers. Built around a survey of the frequency of numbers on webpages.
www.turbulence.org/Works/nums/

Index